Magnetic Oxides and Composites

Edited by
Rajshree B. Jotania
Sami H. Mahmood

Magnetic oxides, their composites and nanoparticles are uniquely suited for a wide variety of applications in new technologies, including device miniaturization, power efficiency improvement and health sector innovations. The interest in these materials is due to such properties as high resistivity, low dielectric and magnetic losses, good corrosion resistance and favorable mechanical characteristics. The book focuses on the relevant basic concepts, as well as on synthesis routes and important applications of spinel ferrites, hexaferrites and magnetic oxide nanomaterials.

Keywords: Magnetic Oxides, Spinel Ferrites, Hexaferrites, Magnetoelectric Ceramic Composites, Soft Ferrites, Nano-Size Spinel Ferrites, Magnetic Nanoparticles, Device Miniaturization

Magnetic Oxides and Composites

Edited by

Rajshree B. Jotania[1]
Sami H. Mahmood[2]

[1] Department of Physics, University school of sciences, Gujarat University,

Ahmedabad 380 009, India

[2] Department of Physics, The University of Jordan, Amman-11942, Jordan

Published by **Materials Research Forum LLC**
Millersville, PA 17551, USA

Published as part of the book series
Materials Research Foundations
Volume 31 (2018)
ISSN 2471-8890 (Print)
ISSN 2471-8904 (Online)

Print ISBN 978-1-945291-68-5
ePDF ISBN 978-1-945291-69-2

Distributed worldwide by

Materials Research Forum LLC
105 Springdale Lane
Millersville, PA 17551
USA
http://www.mrforum.com

Manufactured in the United States of America
10 9 8 7 6 5 4 3 2 1

Table of Contents

Preface

Magnetic oxides with high resistivity, low dielectric and magnetic losses, corrosion resistance, and favourable mechanical characteristics have played an important role on the development and the expansion of the range of applications in emerging technologies. The possibility to tune the magnetic and electrical properties of these oxides over a wide range by adopting a variety of synthesis routes and cationic substitution scenarios, coupled with the ease of production in large quantities at low cost, have led to the dominance of these materials over their metallic counterparts in uncountable number of applications in almost every aspect of our life. As such, potential utilization of the full range from soft to hard magnetic oxides with cubic, hexagonal, or garnet structures in a plethora of technological and industrial applications have driven great investments in the development, production and modification of the properties of these materials. In addition, the usefulness of nanoparticles and composites in device miniaturization, improvement of performance and power efficiency, and potential for applications in health sector and new evolving technologies have made these materials of special interest to scientists, engineers and technologists. The present special topic volume on "Magnetic Oxides and Composites", to be published as part of the book series "Materials Research Foundation", contains a total of nine chapters, focusing on topics related to ferrites and magnetic oxide nano particles. In these chapters, authors attempted to cover basic relevant concepts, synthesis routes and most important applications of spinel ferrites, hexaferrites and magnetic oxide nano materials.

Chapter one by R. C. Pullar *et al.* is concerned with the synthesis and characterization of barium hexaferrite/barium titanate and strontium hexaferrite/barium titanate composites. Structural analysis of composites based on hexaferrites prepared by four different synthesis routes (ceramic, coprecipitation, citrate and sol-gel) were presented, and the magnetic properties of selected composites were discussed. Prepared composites have potential applications as magnetoelectric materials.

In the second chapter, S. H. Mahmood and I. Bsoul reported a brief review on the common experimental techniques and preparation conditions used for the synthesis of M-type hexaferrites and their influence on the magnetic properties. The effects of various cationic substitution scenarios on the magnetic properties of the hexaferrites were revisited in some details. In particular, the effects of Co-Ti substitution on the structural and magnetic properties of SrM hexaferrites were discussed.

Chapter third by M.A. Ahmed *et al.* is focused on the structural and magnetic properties of $Gd_3Fe_5O_{12}$ prepared using auto-combustion method with glycine as fuel. The effect of (glycine/ nitrate) ratio on the structural and magnetic properties was investigated. A remarkable improvement in the densification of the compound with increasing (glycine/nitrate) ratio was reported.

Chapter four by S. H. Mahmood *et al.* is focused on the study of structural and magnetic properties of vanadium substituted SrM, and Europium substituted BaM hexaferrites prepared by high energy ball milling technique and sintering at 1200° C. Phase separation was reported in the V-substituted SrM hexaferrite with a consequent reduction of the magnetic properties. Also, partial substitution of Ba by Eu in BaM hexaferrite led to the evolution of Eu-garnet secondary phase along with pure BaM and α-Fe$_2$O$_3$. Based on the results of the study, the authors proposed a procedure for the production of hexaferrite/garnet composites, and high-quality hexaferrites with potential for high density magnetic recording.

Chapter five by G. Aravind *et al.* is concerned with the electrical and magnetic properties of cobalt substituted lithium nano ferrites synthesized by citrate gel auto combustion route. Single phase cubic spinel ferrites were prepared with an average crystallite size in the range of 36-42 nm, and DC resistivity suggesting semiconducting behaviour. Cobalt substitution was reported to transform the soft pure lithium ferrite into magnetically hard cobalt substituted lithium ferrite. Superparamagnetic behaviour observed in Li-Co ferrites was claimed to be useful in biomedical applications for targeted drug delivery and magnetic resonance imaging (MRI).

The influence of anionic surfactant additions on the structural, microstructural, magnetic and dielectric properties of Sr$_2$Cu$_2$Fe$_{12}$O$_{22}$ hexaferrites was discussed by R. A. Nandotaria *et al.* in chapter six. Multicomponent products composed of mixtures of Y, M and α-Fe$_2$O$_3$ were obtained, and improvement of the surface morphology was observed as a result of surfactant additions. The saturation magnetization of the samples ranged between a maximum value of 45.0 emu/g and a minimum value of 31.47 emu/g; whereas the coercivity was in the range from 75 to 500 Oe. The magnetic analysis confirmed the formation of multi domain structure, and the dielectric properties of all prepared samples revealed normal ferrite behaviour.

B. P. Rao and M. Abbas have contributed chapter seven on "Magnetic Nanoparticles: Fabrication, Properties and Applications". In this contribution, few common and cheap synthesis routes, and characterization of magnetic nano particles of selected spinel ferrites, core shell nanoparticles, and special composites were addressed. The authors presented a brief review of some aspects of high frequency device applications, magnetostrictive applications, and biomedical applications of ferrite nanoparticles.

Structural, magnetic and dielectric properties of Al- Co substituted M−type strontium hexaferrites are reported in chapter eight by Chetna C. Chauhan *et al.* In this chapter, the authors discussed the effect of Co^{+2}–Al^{+3} substitution on structural, morphological, magnetic and dielectric properties of $SrCo_xAl_xFe_{(12-2x)}O_{19}$ ($0.0 \leq x \leq 1.0$) hexaferrites, prepared using a simple heat treatment technique.

A brief review of the structural-magnetic behavior and application of nano size spinel ferrites in chapter nine was contributed by N. N. Sarkar, *et al.* The authors reported on the crystal structure and cation distributions in spinel ferrites. Also, aspects related to superparamagnetism, and magnetic behaviour were discussed. In addition, some applications of spinel ferrites were included in this chapter

The editors would like to thank all authors for their valuable contributions, to referees (Vaishali Soman, K. M. Jadhav, Ibrahim Bsoul, Ibrahim Abu-Aljarayesh, and others) for their critical comments, and to Materials Research Forum LLC, U.S.A. for publishing this special topic volume in the form of a book.

Sami H. Mahmood

Rajshree B. Jotania

Chapter 1

Synthesis and Characterisation of Magnetoelectric Ceramic Composites based on M-type Strontium and Barium Hexagonal Ferrites and Barium Titanate

Marco S. A. Medeiros[1,a], João S. Amaral[1,b], and Robert C. Pullar[1,c]

[1] Department of Materials and Ceramic Engineering / Department of Physics / CICECO - Aveiro Institute of Materials, University of Aveiro, Campus Universitário de Santiago, 3810-193 Aveiro, Portugal

[a]marcosamedeiros@gmail.com, [b]jamaral@ua.pt, [c]rpullar@ua.pt

Abstract

Magnetoelectric composite ceramics were prepared to study their phase compatibility, magnetic and piezoelectric/ferroelectric properties, and coupling between magnetic and ferroelectric properties. The synthesis of various $BaFe_{12}O_{19}$ and $SrFe_{12}O_{19}$ hexaferrites was undertaken with different sintering temperatures, and exploring four different methods: solid state reaction, coprecipitation, sol-gel and citrate (Pechini) routes. Ceramic composites of BaM and SrM with $BaTiO_3$ (BT) as a ferroelectric/piezoelectric phase, were prepared with both uniaxial and isostatic pressing, and then sintered. The composites were characterised by XRD, SEM and VSM. Results showed that BaM and BT did not react in the composites, while SrM-BT composites possess SrM, BT and $SrTiO_3$ phases.

Keywords

Magnetoelectric Composite Ceramics, Structural Behavior, SEM, Magnetic Properties, Hexagonal Ferrites

Contents

1. Introduction

1.1 Magnetoelectric materials

Magnetoelectric coupling was first investigated at the end of the 19th century, when the polarisation of a dielectric under the influence of a magnetic field was first postulated by Röntgen [1], and in 1894 Pierre Curie pointed out that it would be possible for an asymmetric molecular body to polarise directionally under the influence of a magnetic field [2]. The actual term "magnetoelectric" is attributed to Debye in 1926 [3], after the first unsuccessful attempt to demonstrate the static magnetoelectric effect experimentally [4], and in 1948 Tellegen formulated the idea to fabricate a composite displaying magnetoelectric effects [5]. However, it was only in 1959 that Dzyaloshinskii predicted the magnetoelectric effect in Cr_2O_3 based on symmetry considerations, and in 1960 Astrov confirmed this prediction experimentally by measuring electric field induced magnetisation [6-8].

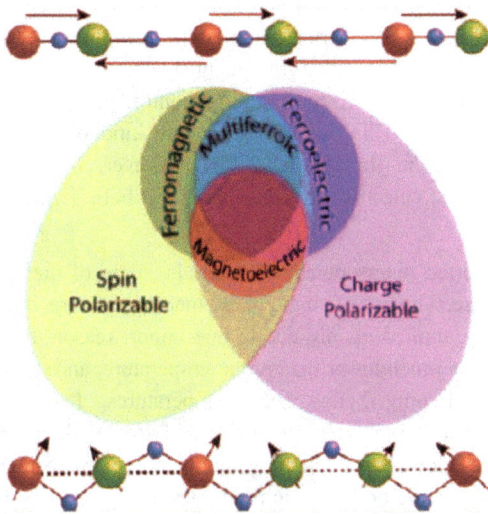

Fig.1. Diagram of the multiferroic and magnetoelectric properties relationships.

The first attempt to combine two different proprieties, ferromagnetic and ferroelectric in the same system was in the 1960's by two groups in the Soviet Union, the group of Smolenskii [9] and by Venevtsev [10]. Novel materials were developed with both ferromagnetic and ferroelastic/ferroelectric properties, and the name multiferroic was coined for this kind of material [11]. Multiferroics combine at least two ferroic orders - ferroelectric, ferromagnetic, or ferroelastic - in the same temperature range. These materials can be single-phase, which are rare but do exist, or composites [12]. The ferroic systems possess a parameter order that is switchable by an adequate driving force or field, phenomena normally accompanied by hysteresis. Magnetoelectricity can also involve coupling of non-ferroic properties, such as permeability, permittivity or resonant frequencies, and be done by direct stimulus due to an applied field, or an indirect stimulus through the application of voltage or stress. The relationship between these is shown in Fig. 1

The development of new types of materials has been the key for the innovation of technology in the present day, such the development of required materials with high multifunctionality and low dimensions. Magnetoelectric ceramics are considered by many researchers to be a good choice to produce new kinds of devices. They can exhibit ferroelectric and magnetic properties, and some coupling between magnetic and dielectric/ferroelectric phases, called the magnetoelectric effect, where an induced electrical polarisation or magnetisation can be controlled by applying a magnetic or electric field, respectively. This would allow some kind of tuneability, switching or sensing to be achieved in these composites. However, in order to be useful for applications, the magnetoelectric coupling must be both large and active at room temperature.

There are many reported magnetoelectric materials, some of them are composites and others are single phase. The use of single phase materials is rare, and most of them have not been successfully applied mainly due to one import reason, they have low Neel or Curie temperature (T_c), much lower than room temperature, and those materials exhibit a magnetoelectric effect only below such temperatures. Furthermore, the intrinsic structural/electronic conditions which give rise to ferroelectricity are not conducive to ferromagnetism, and vice-versa, rendering single phase magnetoelectric materials as having either poor magnetic or dielectric properties, and exhibiting low magnetoelectric coupling.

1.2   Magnetoelectric composites

An alternative approach is to make a composite material, consisting of two complimentary magnetic and ferroelectric phases, which can have coupling between them, often strain-mediated in nature (i.e., piezoelectric/magnetostrictive coupling). As an alternative to overcoming the problems of the single-phase magnetoelectric materials and with greater design flexibility, magnetoelectric composites made by combining piezoelectric and magnetic materials together have become much more attractive and gained significant interest in the last few years due to their multifunctionality, because the coupling interaction between piezoelectric and magnetic materials can produce a large magnetoelectric response [13]. This response is potentially several orders of magnitude higher than that in current single-phase magnetoelectric materials at room temperature.

A new class of multifunctional devices could be built from magnetoelectric composites such as magneticelectric transducers, actuators and sensors. The magnetoelectric effect in composite materials is known as a product tensor property resulting from the cross interaction between different orderings of the phases in the composite. The individual phases don't have magnetoelectric effects, but in composites the two phases (piezoelectric and magnetic) can have a remarkable combined magnetoelectric effect [14].

In composite materials, the ME effect was first realised after Van Suchetelene introduced the concept of "product properties" in 1972. He used it to successfully grow the first magnetoelectric composite by unidirectional solidification of a $BaTiO_3$–$CoFe_2O_4$ eutectic liquid [15]. In 1978 Van den Boomgaard and Born [16] postulated that ideally: (i) the two individual phases should be in equilibrium; (ii) mismatching between grains should not be present; (iii) the magnitude of the magnetostriction coefficient of the piezomagnetic / magnetostrictive phase, and the magnitude of the piezoelectric coefficient of the piezoelectric phase, should be of a similar order and as large as possible; (iv) accumulated charge must not leak through the piezomagnetic or magnetostrictive phase; and (v) there should be a deterministic strategy for any electrical poling of the composites.

In the early 1990's, Newnham's group and Russian scientists prepared ceramic composites by conventional sintering methods, consisting of spinel ferrites as the magnetic phase and $BaTiO_3$ (BT) or $Pb(ZrTi)O_3$ (PZT) as the ferroelectric/piezoelectric phase [14]. In the beginning of the 21st Century magnetoelectric composites materials started to gain attention, and an upsurge in the magnetoelectric composite research occurred in the early 2000s. Ceramic composites with different connectivity schemes were reported, such as the 0–3 type and 2–2 type laminate composites [14], although by far the easiest composite form to synthesise are simple 0-0 particulate composites, and that is where the most advances have been made [17].

Magnetoelectric composites can exhibit magnetoelectric effects over a wide temperature range [18]. This effect can be archived by different paths, such as using the resulting propriety of the piezoelectric and magnetostrictive effects [19, 20], or using the product propriety of pyroelectric and magnetostrictive effects [17]. The composite materials properties of magnetostrictive and piezoelectric materials can be explained by the following mechanism: when a magnetic field is applied to the composite the magnetostrictive material is mediated/strained, and this will induce a stress on the piezoelectric component and generate an electric field. The inverse of this effect is also possible - applying an electrical field to the piezoelectric material will produce strain, which will be transferred as stress to the magnetostrictive material, producing a change in the magnetic permeability of the material. In both cases the product property resulting in such composites is described as the magnetoelectric effect, in which a magnetic applied field induces a change in electric polarisation, and an applied electric field induces changes in magnetic permeability in the composite. Therefore, the ME effect is the result of the product of the magnetostrictive effect (magnetic/mechanical) in the magnetic phase and piezoelectric effect (mechanical/electrical) in piezoelectric phase. This effect can be quantified in two ways:

$$ME \; effect = \frac{electrical}{mechanical} * \frac{mechanical}{magnetic} units \; of \; \alpha = s \, m^{-1}$$

$$ME_{II} \; effect = \frac{magnetic}{mechanical} * \frac{mechanical}{electrical} units \; of \; \alpha_{II} = s \, m^{-1}$$

(1)

The magnetoelectric (ME) effect is when an electric field is applied, and the inverse magnetoelectric effect (ME_{II}) is when a magnetic field is applied [21].

1.3   Phase connectivity in magnetoelectric composites

Fig.2. The ten different levels of connectivity that can exist in composites, as defined by Newnham [25].

There are various ways in which the components can be connected in a magnetoelectric composite material, such as the bulk ceramic magnetoelectric composites of piezoelectric ceramics and ferrites [14, 22-24], two-phase magnetoelectric composites of magnetic alloys and piezoelectric materials, and layered thin films of ferroelectric and magnetic oxides. In 1978, Newham *et al.* introduced the concept of phase connectivity [17]. With this concept it is possible to describe the structure of two phase composites, where each phase in a composite may be self-connected by different dimensions zero, one, two and three dimensions.

In the case of two-phase composites, there are ten different connectivities: 0-0 (zero connectivity between either of the phases present, i.e. a random mix of phases), through to 3-3 (both phases interpenetrating in all three dimensions).The different changes in connectivity between phases can result in substantially different proprieties, and Fig. 2 illustrates the different connectivities.

Fig. 3. SEM image of cellular hexaferrite ceramics made from biomorphic cork templates.

3-3 phase connectivity is the most complicated to obtain, and potentially the most interesting one, with the two phases of the composite forming interpenetrating three-dimensional networks. One way to achieve tailorable 3–3 composites is infiltration of a second phase into porous materials that display complete pore connectivity and percolation. Provided the structure of the initial porous material can be precisely controlled in terms of the degree of porosity, the size and shape of the pores, and the nature of the struts separating them, then there is the opportunity to design and fabricate interpenetrating composites with customised structures. Hence, the infiltration of cellular ceramics offers the potential for producing tailored ceramic-based interpenetrating composites with 3–3 interconnectivity. Patterns of this phase connectivity occur in some living systems such as corals where organic tissue and an inorganic skeleton interpenetrate one into another. Another possibility could be impregnation of wood/plant template cellular materials, such as the cork-derived ferrite ceramics (Fig.3) developed by R. C. Pullar *et al.* [26, 27], with a second ferroelectric/dielectric phase.

1.4   Domains, magnetostriction, ferroelectricity and piezoelectricity

In magnetic materials the magnetic moments of the electrons become aligned, forming regions where they show spontaneous magnetism, and these regions are called domains. In a non-magnetised sample of a magnetic material the domains are randomly distributed, and the total magnetic field in any direction is zero, as shown in Fig. 4. When an external magnetic field is applied to magnetic materials, the domains aligned to this field grow at the expense of the other, unaligned domains, and if a sufficiently strong external field is applied, all of the domains will change to the direction of the external field (Fig. 4).

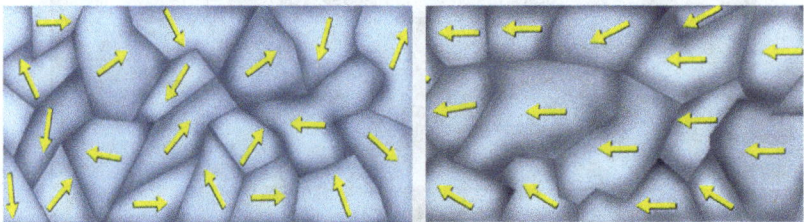

Fig.4. Randomly oriented (left) and aligned (right) domains.

When all the domains have realigned along the field direction, the magnetisation of the material cannot increase further, and it is said to have reached magnetic saturation (M_s). When the external magnetic field is removed, the alignment of the domains may decrease, along with the magnetisation, but a ferromagnetic material will remain magnetised, maintaining a net magnetisation after the field is removed, and this is called the remnant magnetisation (M_r). To demagnetise the material, and make all the domains randomly oriented again, the material needs to be either heated above its Curie temperature (T_c), or a magnetic field of sufficient magnitude has to be applied in the opposite direction. The magnetic field required to "coerce" the domains into a random state is the coercivity (H_c), and when a large field is needed, the material is said to be a hard magnet.

Some materials can change shape when in the presence of an external magnetic field, and this effect is called "Magnetostriction". Magnetostrictive materials are able to convert magnetic energy into mechanical energy (or the reverse), and this effect occurs in most ferrimagnetic materials. This phenomenon is attributed to the rotations of small magnetic domains in the material, which are randomly oriented when the material is not exposed to a magnetic field. The orientation of these small domains by the imposition of the magnetic field creates a strain field, and produces a physical length change in the material. As the intensity of the magnetic field is increased, more and more magnetic domains orient themselves so that their principal axes of anisotropy are collinear with the magnetic field in each region, and finally a saturation is achieved. Conversely, applying a stress which causes the domains to align can generate magnetisation with the material.

The electrical equivalent of this is a ferroelectric material, in which areas of the molecule have concentrated positive or negative charge, creating an electric dipole, which can be aligned along an electrical field. Normally these form ferroelectric domains which are randomly oriented throughout a material, but they can also be aligned by applying an electrical field, and reversed to zero or an opposite polarisation by applying an opposite electrical field. This creates a ferroelectric hysteresis loop of electrical polarisation vs. applied field in ferroelectric materials, similar to that for magnetisation observed in ferrimagnetic (or ferromagnetic) materials. Analogous to magnetostriction, such materials also exhibit piezoelectricity, that is when a stress is applied to deform and align the ferroelectric domains, a net charge can be created causing a current to flow. Conversely, applying an electrical field can align the domains, creating strain as they deform and changing the dimensions of the material. When the field is removed, a degree of strain usually remains, and this is called "poling" a piezoelectric material. A reverse field is required to remove this strain, so there is also a piezoelectric hysteresis loop of strain vs. electrical field.

It is the coupling of these ferrimagnetic and ferroelectric/piezoelectric domains that enables a composite to become magnetoelectric, where an applied magnetic field can vary electrical properties, or an applied electrical field can change the magnetic properties.

2.   Hexagonal ferrites and barium titanate

Usually ferrites are primarily classified into three types: garnets, spinels and hexagonal ferrites, according to their primary crystal lattice. Generally, ferrimagnetism arises from the antiparallel alignment of the magnetic moments on transition metal ions, present on different magnetic sublattices. The origin of the antiparallel coupling can be explained by the superexchange of valence electrons between the filled p- orbital of O^{2-} and unfilled d-orbital of the transition metal cations. In ferrites, the oppositely directed magnetic moments do not exactly cancel, thus a net magnetic moment results [28, 29]. This is known as ferrimagnetism, in which there are opposing spins in opposite directions, but there are more spins in one direction, giving a net overall magnetisation (Fig. 5).

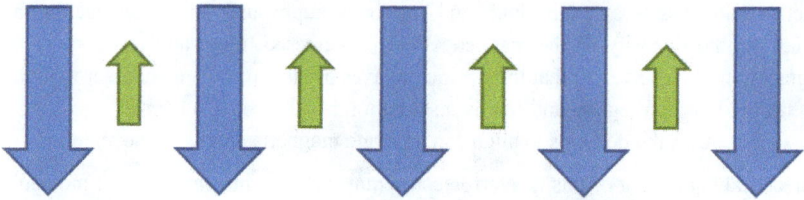

Fig. 5. Opposing, but unequal, magnetic spins in a ferrimagnetism material resulting in a net magnetisation.

2.1   Hexagonal M-ferrites, $BaFe_{12}O_{19}$ and $SrFe_{12}O_{19}$

The hexagonal ferrites are a group of ferrites with hexagonal structure, and they are all ferrimagnetic materials, with their magnetic properties intrinsically linked to their crystalline structures [24]. The hexaferrites have a directional preference of magnetisation called magnetic anisotropy, and in the M ferrites, such as $BaFe_{12}O_{19}$ (BaM), this is aligned along the c-axis of the hexagonal lattice (Fig. 6). This "locks" the magnetisation into this direction, making them magnetically hard, which is used to good advantage in many applications as a permanent magnet material [30]. Much more detail about the structure, chemistry and properties of the hexagonal ferrites is given in the review by Pullar [24].

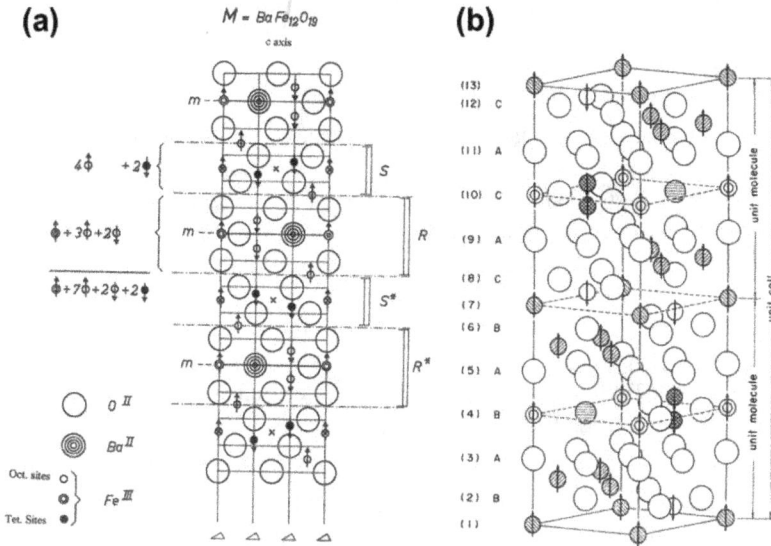

Fig. 6. Cross section view (a) of the M ferrite (BaFe$_{12}$O$_{19}$) structure, in which the vertical lines are axes of threefold symmetry along the c-axis [28], and (b) perspective views of the M unit cell [31].

The magnetic mineral magnetoplumbite was first characterised in 1925 [32], and in 1938 the crystal structural was found to be hexagonal with the composition PbFe$_{7.5}$Mn$_{3.5}$Al$_{0.5}$Ti$_{0.5}$O$_{19}$ [33]. The synthetic form of magnetoplumbite was found to be PbFe$_{12}$O$_{19}$, or pure PbM, and a number of isomorphous compounds were suggested including BaFe$_{12}$O$_{19}$. However, it was only after the Second World War that this material was structurally investigated by Philips Laboratories, and in the 1950's the development of the hexaferrites started, with Went et al. reporting that BaM has a hexagonal structure [34].

After this, hard magnets such as BaFe$_{12}$O$_{19}$and SrFe$_{12}$O$_{19}$ experienced considerable development, and the most common M ferrites are still BaM and SrM (SrFe$_{12}$O$_{19}$), both of which are used on a very large scale as permanent magnets [35]. BaM is a hard magnet with a maximum coercivity of over 500 kA m^{-1} (although 300-400 kA m^{-1} is more common in polycrystalline ceramics), saturation magnetisation of around 72 A m^2 kg^{-1}, and high electrical resistivity of $10^8 \Omega$ cm [24]. In comparison with many magnetic alloys, BaM has a lower saturation magnetisation but has high magnetic uniaxial anisotropy along the c-axis [28], and useful microwave properties.The maximum theoretical density of BaM is 5.295 g cm^{-3}[36].

2.2   Barium titanate, BaTiO$_3$

Since it was discovered, Barium titanate (BaTiO$_3$) has been the object of a vast number of studies, due to its attractive proprieties. BaTiO$_3$ is a ferroelectric material with good electrical proprieties at room temperature, and is very stable chemically and mechanically. It can be easily prepared and used in the form of ceramic polycrystalline samples. BaTiO$_3$ shows high dielectric constant and dielectric low loss characteristics up to microwave frequencies, and these proprieties are useful to make devices such as capacitors and multilayer capacitors (MLCs). When doped with other elements such strontium or lanthanum, BaTiO$_3$ can also be used in more wide-spread applications, such as semiconductors, PTC thermistors, dielectric bolometers, infrared sensors and others [37]. It has a molecular mass of 223.192 g mol^{-1} and a density of 6.02 g cm^{-3}.

BaTiO$_3$ belongs to the large family of compounds called perovskites, named from the mineral perovskite (CaTiO$_3$) characterised by Gustav Rose in 1839 [38]. The stoichiometry of the perovskite structure is ABO$_3$, and the ideal perovskite structure adopts the cubic space group P$_m3_m$ with a face-centered cubic (fcc) structure, defined by the cations A and B, where A is a large divalent cation (i.e., Ba) situated in the corners of the cube, B is a smaller tetravalent cation (i.e., Ti) in the middle of the cube, and O is an anion, commonly oxygen, in the centre of the face edges, as described in Fig. 7.When the cation A possess a large ionic radius, as in the case of Pb, Ba or Sr, the structure can become distorted, which results in polarisation [39].

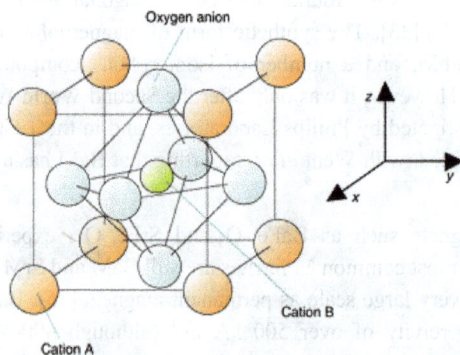

Fig.7. Crystal structure of cubic ABO$_3$perovskite [40].

Therefore, many perovskites, such as $BaTiO_3$ (BT), have different crystal structures other than cubic at lower temperatures. In the case of BT, it has four common polymorphs. From high temperatures of 1460°C down to the Curie temperature (T_c) of ~130°C (for polycrystalline BT), the crystal structure is cubic, with the central titanium ion surrounded equally by oxygen ions (O^{2-}) on each of the cube faces, and a barium ion on each corner of the unit cell, as shown in Fig.8. In this cubic crystal structure there is no electrical dipole moment, and hence it has no ferroelectric or piezoelectric properties [38, 39]. However, below the T_c, BT adopts a tetragonal structure (Fig. 8), in which the atoms of titanium and barium are dislocated resulting in a slight distortion of the structure. At this temperature the crystallographic structure passes from the cubic paraelectric phase to a ferroelectric tetragonal structure. With further decreases in temperature other phases occur, depending on the temperature, as shown in the diagram below [41]. These lower temperature phases are all ferroelectric, as the non-symmetrical displacement of the central Ti atoms in at least one axis results in a net polarisation of charge within the crystal.

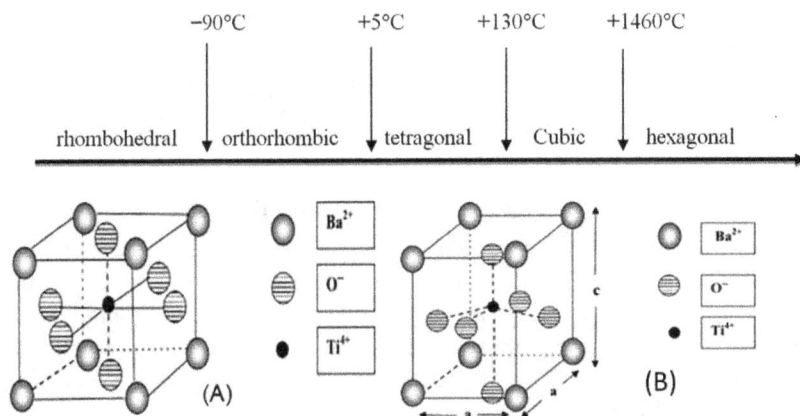

Fig. 8. Cubic (m3m) crystal structure of $BaTiO_3$ above T_c (A), Tetragonal crystal structure of $BaTiO_3$ at room temperature (B) [38, 39].

The titanium ions and the octahedral arrangement of the oxygen ions displace asymmetrically, causing a permanent electric dipole moment in the unit cell. At this temperature range the crystal structure is stable and shows ferroelectric proprieties along the direction <001> and a high dielectric constant which is ideal for capacitors [41].

For temperatures below 5°C the ferroelectric proprieties are maintained by changing the direction of the polarisation to firstly the <110> orthorhombic crystal structure and then to the direction <111> in the rhombohedral structure[42].

3.   Materials and composites synthesis

The chemicals used in this work to synthesise the samples from all the different techniques had high purity and were bought from well-known suppliers. Specifically, the chemicals used were:

Iron (III) oxide (Fe_2O_3), BDH Prolabo, A.C.S reagent 99%, MW =159.69 g

Barium carbonate ($BaCO_3$), Sigma-Aldrich, 99+%, MW = 197.32 g

Strontium carbonate ($SrCO_3$), Sigma-Aldrich, 98+%, MW = 147.63 g

Titanium (IV) oxide (TiO_2), Sigma-Aldrich, Reagent Plus 99+%, MW = 79.87 g

Iron III nitrate nonahydrate ($Fe(NO_3)_3 \cdot 9H_2O$), Sigma-Aldrich, A.C.S reagent 98+%, 404.00 g

Barium nitrate ($Ba(NO_3)_2$), Sigma-Aldrich,A.C.S reagent 99%, MW = 261.34 g

Strontium nitrate ($Sr(NO_3)_2$), Sigma-Aldrich,A.C.S reagent 99%, MW = 211.63 g

Citric acid monohydrate ($C_6H_8O_7 \cdot H_2O$), Sigma-Aldrich, A.C.S reagent 99.5%, MW = 210.14 g

Ammonia solution (NH_3/NH_4OH), Merck, 25% solution, MW = 35.05 g

Barium titanate was prepared by heating a mixture of barium carbonate ($BaCO_3$) and titanium (IV) oxide (TiO_2). The compounds were weighed and then ground in a mortar until well mixed, and ball milled with 10 mm zirconia balls in isopropanol for 12 h for further mixing. This fine powder was dried in an oven until the complete evaporation of isopropanol had occurred, and the mixture was then calcined in a furnace at the temperature of 1350°C for 10 h in air.

For synthesis of the hexaferrites, four different synthesis methods were compared:

- Standard solid state powder reactions
- Coprecipitation
- Sol-gel
- Citrate combustion (Pechini-type) synthesis

3.1 Solid state reaction route for hexaferrites

The M-hexaferrites (barium, strontium) were prepared from a mixture of carbonates ($BaCO_3$ or $SrCO_3$) and Fe_2O_3 oxide, in the stoichiometric ratio of Ba/Sr to Fe_2O_3 of 1:6. After properly mixing with mechanical stirring for 10 min in isopropanol, the powders were put in a Teflon pot with zirconia balls, and then placed on a rotary ball mill and left to mix for 12 h. After mixing, the pot was dried for 48 h, and the resulting powder was then manually mixed in a mortar until the mixture was homogenous. This was then calcined at different temperatures of 1000°C, 1050°C, 1100°C, 1150°C and 1200°C for 2 h at heating rates of 5°C/min in air.

3.2 Coprecipitation synthesis of hexaferrites

The chemical method of coprecipitation of salts with a base results in the formation of an insoluble precipitate (usually a hydroxide) from aqueous solutions of nitrates, chlorides or sulphates of Fe^{3+}, and of divalent cations such as Ba, Sr, Co, etc., resulting in a precipitate containing all the components mixed at an atomic level. This wet chemical process improves homogeneity. In this process a strong base such as sodium hydroxide (NaOH) or ammonia (NH_3) is added to precipitate the solution. To avoid problems when precipitating, a pH higher than 10 is needed, because of the different solubility dependences of some metal hydroxides and metal cations in the solution [24, 43].

The hexaferrites were made by this process from nitrate salts $Ba(NO_3)_2$ or $Sr(NO_3)_2$, and $Fe(NO_3)_3 \cdot 9H_2O$, in stoichiometric ratios of 1:12 for Ba/Sr:Fe nitrate. Each composition was mixed by mechanical stirring in a large beaker with distilled water, the nitrates dissolved in the water within 30 min. Then a total of 300 ml of a 5% ammonia solution was added dropwise with constant mechanical stirring, causing a precipitate to form. A pH of 9.3 was obtained in the solutions, and then the suspension was filtered, the solid part washed with distilled water to remove any remaining nitrate and ammonium ions, and the solid cake obtained was dried in an oven for 24 h. This was then calcined at different temperatures, as in the solid state reaction route above.

3.3 Sol-gel synthesis of hexaferrites

Sol-gel is a wet-chemical technique widely used in materials science to obtain ceramic materials, typically metal oxides. The chemical process of sol-gel is based in reactions of inorganic polymerisation/peptisation of metal hydroxides or metallo-organics on the colloidal scale. In general, the sol-gel process involves the transition of a solution system from a liquid "sol" (a colloidal suspension) into a solid "gel" phase, with intimate mixing of the components on the near-atomic level. Utilising the sol-gel process, it is possible to fabricate advanced materials in a wide variety of forms: ultrafine or spherical shaped powders, thin film coatings, fibres [44], porous or dense materials, and extremely porous aerogel materials.

A sol is a colloidal suspension of solid particles in a liquid were the gravity forces are cancelled by short range forces such as Van der Walls, surface charges and Brownian motion, and the particles within the sol typically have a diameter of 1 to 1000 nm. Sols are formed through a hydrolysis/ condensation reaction of colloidally-sized particles [24]. The hydrolysis happens when the metallic precursor is dissolved by water, and then the metallic cations are hydrolysed by the water molecules and forms metal hydroxides, as shown in the equation:

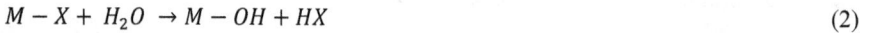

$$M - X + H_2O \rightarrow M - OH + HX \tag{2}$$

Where the M is a metal and X is a reactive species such as nitrates. The reaction starts when the OH group is added, that reacts with the metal ion in the solution, and can be acid or base catalysed. Condensation occurs by the linking together of hydrolysed molecules to form M-O-M connections, with the loss of a molecule of water:

$$M - OH + HO - M \rightarrow M - O - M + H_2O \tag{3}$$

There are many factors that influence the reactions in hydrolysis and condensation, such as pH, temperature, metal ions concentration [45]. When the sol starts to concentrate or condensation completes, it forms a continuous amorphous connected structure knows as a gel.

To produce ferrites by aqueous sol-gel synthesis, an aqueous solution of metal ions is coprecipitated by a base, but then instead of drying and firing the precipitates, they are treated with an acid catalyst to form a colloidal sol, which can then be concentrated to a gel and then can be dried to make powder and fired to give the ferrite, as in the process developed by Pullar *et al.* to create BaM and SrM ferrite sols and fibres from inorganic salts [46-49].

In this sol-gel method, the samples were made as in the coprecipitation method, but instead of drying and firing the precipitate, 3.5ml of concentrated nitric acid (as a 5% solution in 20 ml of distilled water) was added to it, and this was then digested in a 500 ml evaporating flask attached to a rotary evaporator (Rotavapor® R-210/R-215, Buchi) for 24 h at a temperature of 60°C while slowly evaporating. This enabled the hydroxides to peptise, and a colloidal suspension (sol) of hydroxides was obtained. This was then dried completely in an oven for 48h at a temperature of 85°C, and the gel obtained was calcined at different temperatures as in the solid state reaction route above.

3.4 Citrates synthesis of hexaferrites

The citrate method constitutes a polymeric approach, in which different metals are confined in a gelified citric acid network. In this method ultrafine particles can be synthesised at low temperatures, through a violent exothermic reaction from the decomposition of the citrates, in which CO_2 is generated producing a very porous product with a high surface area. It is sometimes also known as the Pechini method [50]. In this process, the citric acid acts as a complexing agent forming metal-citrato complexes, and the metal precursors used are normally nitrates with water as solvent. In the subsequent, highly exothermic, combustion reaction, the citrate acts as the fuel and the nitrate as the oxidiser, and the intense localised heat results in the formation of the desired crystalline phases, without the large grain growth normally associated with heating a sample to the temperatures usually required to achieve this. Often nanoparticles can be obtained by this method.

To prepare BaM and SrM, a stoichiometric solution of nitrates ($Ba(NO_3)_2/Sr(NO_3)_2$: $Fe(NO_3)_3 \cdot 9H_2O = 1 : 12$) was mixed with a magnetic stirrer in distilled water until all was dissolved. Citric acid was then added, in a ratio of total metal cations : citrate of 1. The solution was heated to 80°C, and ammonia was added until the pH reached 7.0 to form a homogenous solution. The solution was then placed on a rotary evaporator (Rotavapor® R-210/R-215, Buchi) to evaporate most of the water, and the rest of the water was removed by drying in an oven for 72 h. After complete drying, the gel was heated to a temperature of 200°C in air, and a mixture of oxide compounds was formed in the subsequent exothermic reaction, which may contain some ready-formed hexaferrites. These powders were then calcined as above.

3.5 Composite synthesis

The ceramic composites (BaM-BT/SrM-BT) were made by mixing each of the hexaferrites (BaM or SrM) from each process described above with the $BaTiO_3$, in the following ratios in weight percentages:

Hexaferrites 90wt% - $BaTiO_3$ 10wt%
Hexaferrites 80wt% - $BaTiO_3$ 20wt%
Hexaferrites 70wt% - $BaTiO_3$ 30wt%

After mixing well in a mortar, pellets were prepared in a uniaxial press with a die diameter of 10 mm and a pressure of 100 MPa. For each composite composition, hexaferrite calcination temperature and hexaferrite synthesis method, three samples were prepared, for a total of 540 samples in all from uniaxial pressing. Some of these samples were given further cold isotactic pressing (cip) under the pressure of 200 Mpa for 5 min.

After pressing, the composite pellets were weighed, and sintered at 1200°C for 2 h in air. After sintering, the density of the pellets was obtained from geometrical measurements of the weight, diameter and height, using the equation below:

$$(\rho) = \frac{m}{V} \tag{4}$$

Where 'ρ' is density of the material, 'm' the mass (g) and 'V' is the volume (cm^3).

3.6 Characterisation and analysis of sintered hexagonal M-ferrites and composites

X-ray powder diffraction (XRD) patterns were recorded at 300 K on a Rigaku Geigerflex D/max-Series instrument (Cu-K$_\alpha$ radiation- λ = 1.5405 Å, 20–70°, 0.02°, 2θ step-scan and 10 s/step), and phase identification by PANalytical X'Pert High Score Plus software and Jade 9.0 software program from MDI. Scanning Electron Microscopy (SEM) was carried out on a Hitachi S-4100 at 25 kV on unpolished samples coated with carbon. The magnetic properties were measured using a vibrating sample magnetometer (VSM, Cryogenics), at 300 K and magnetic fields up to 3 T. The sample temperature was stabilised at 300 K, and the maximum field ramp rate was 0.25 T / min. The device sensitivity is rated up to 10.6 A m^2 kg^{-1}, and the maximum recorded temperature drift was ± 0.05 K, with typical values of ± 0.01 K.

4. Results and discussion

4.1 XRD analysis of M-hexaferrites prepared by diffrent routes

The following figures (Fig.9 to Fig. 24) show XRD patterns of the different hexaferrite (BaM, SrM) powders calcined at different temperatures. As we intended to co-sinter the composites at 1200°C, we did not calcine BaM/SrM hexaferrites powder above 1200°C either.

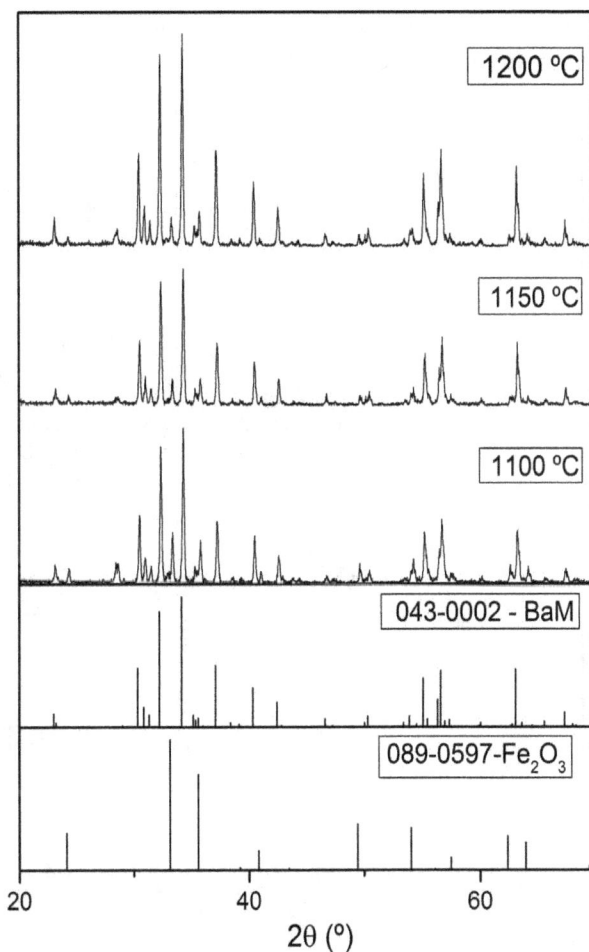

Fig. 9. BaM hexaferrites prepared using solid state reaction and calcined at 1100°C, 1150°C and 1200°C.

For BaM hexaferrites prepared using solid state synthesis and sintered at 1100°C, 1150°C and 1200°C, XRD analysis revealed a good formation of M-ferrite, although there was always the presence of some α- Fe_2O_3 (hematite) in all samples. It can be seen that the formation of BaM starts at 1100°C (Fig. 9), and with the increase of the temperature the quantity of hematite decreases, although BaM is always present as major phase.

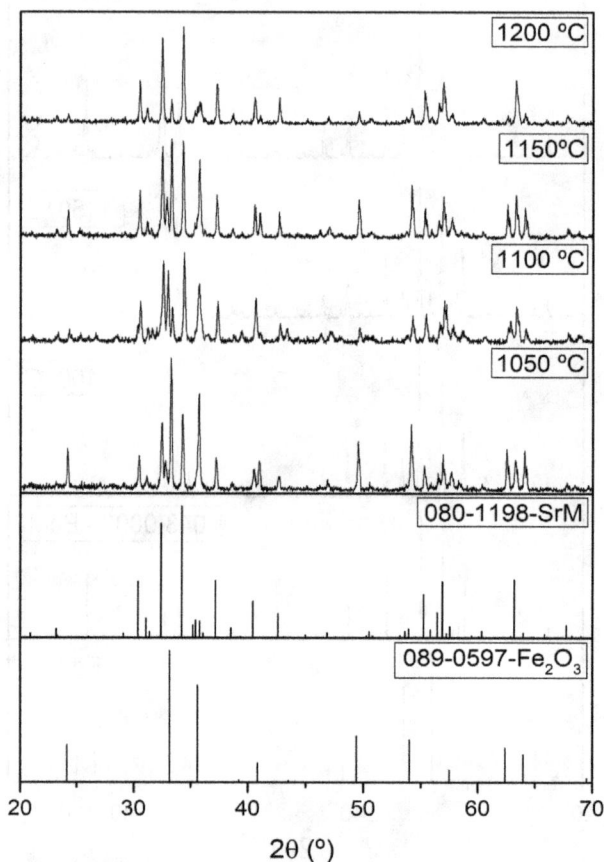

Fig. 10. SrM hexaferrites prepared using solid state reaction and calcined at 1050°C, 1100°C, 1150°C and 1200°C.

XRD patterns of SrM hexaferrites prepared using solid state reaction and calcined at 1050°C, 1100°C, 1150°C and 1200°C are shown in Fig.10. It was observed that BaM constituted < 90% of the phases at 1150 or 1200°C. SrM ferrite formation started at a slightly lower temperature of 1050°C (Fig. 10), as expected, but it was still mostly hematite at this temperature. The phases were approximately equal in quantity at 1100°C, but at 1150°C and 1200°C the samples consisted of ~90% SrM. No M-phase was detected in the XRD patterns for BaM below 1100°C or for SrM at 1000°C, so these are not shown in Figs. 9 and 10.

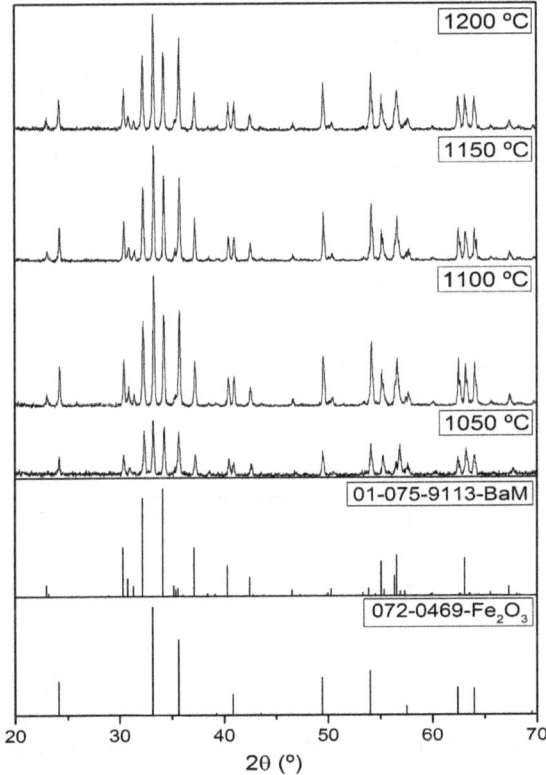

Fig. 11. BaM hexaferrites prepared using coprecipitation route and calcined at 1050°C, 1100°C, 1150°C and 1200°C.

The calcination of M-hexaferrites made by the method of coprecipitation resulted in a much greater amount of secondary phase, with hematite seemingly the major phase at all temperatures (1050°C, 1100°C, 1150°C and 1200°C), for both BaM (Fig. 11) and SrM (Fig. 12). No M-phase was seen for BaM or SrM prepared at 1000°C.

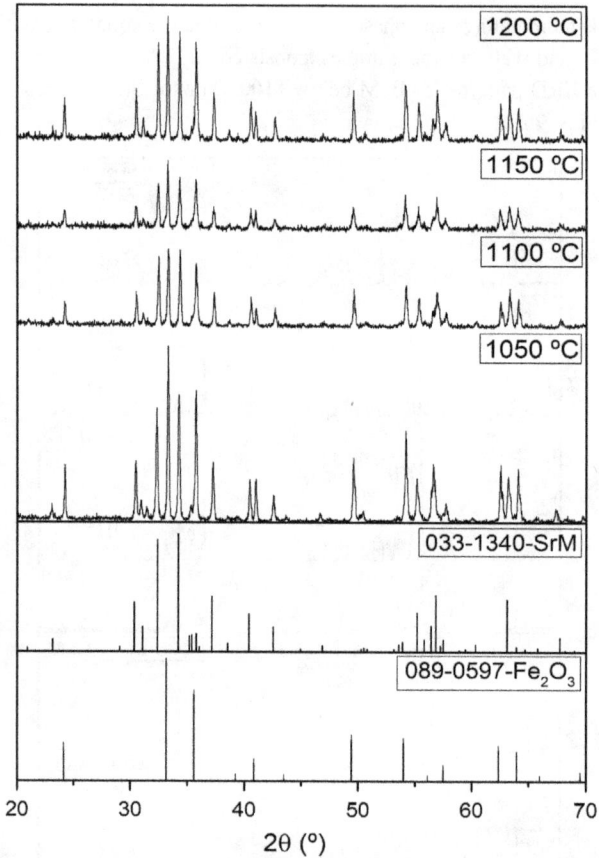

Fig.12. SrM hexaferrites prepared using coprecipitation route and calcined at 1050°C, 1100°C, 1150°C and 1200°C.

Fig. 13 represents BaM hexaferrites prepared using citrate synthesis and calcined at 1000°C, 1050°C, 1100°C, 1150°C and 1200°C, while Fig. 14 shows XRD patterns of SrM hexaferrites prepared using citrate synthesis and calcined at 1000°C, 1050°C, 1100°C, 1150°C and 1200°C. It is interesting to note that with the citrate synthesis method, we achieved the best results for M-Ferrite (Figs.13 and 14), encountering the single M-phase for all the calcination temperatures from 1000°C and upwards. Note that there may be a minute trace of hematite in the samples prepared at 1000°C. The phases had the intensity peaks well defined at all temperatures. Therefore, from the point of view of making a pure M ferrite phase, especially at low temperatures, the citrate synthesis appears to give a considerable advantage. This is probably because a significant amount of nanoscale M ferrite has already crystallised during the combustion, and acts as a template / nucleation site for the rapid conversion of the remaining powder into the M-phase, efficiently and at lower temperatures.

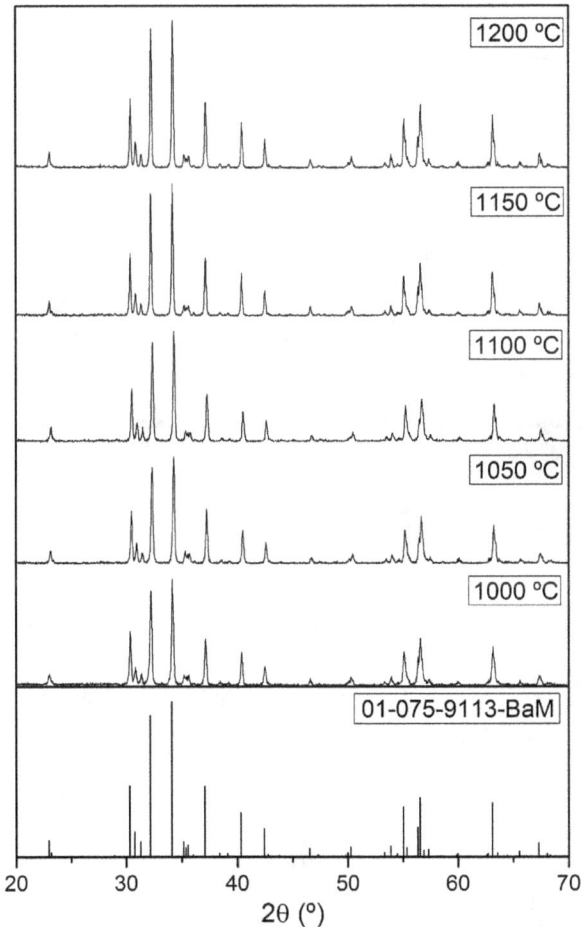

Fig.13. BaM hexaferrites prepared using citrate synthesis and calcined at 1000°C, 1050°C, 1100°C, 1150°C and 1200°C.

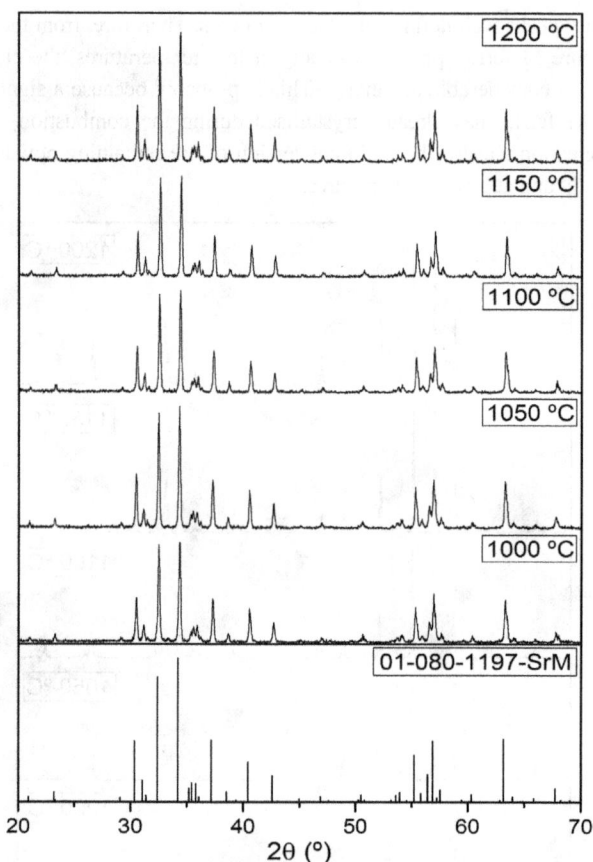

Fig.14. SrM hexaferrites prepared using citrate synthesis and calcined at 1000°C, 1050°C, 1100°C, 1150°C and 1200°C.

The XRD pattern of hexagonal ferrites prepared from the sol-gel route showed dual phases of M-ferrite and hematite in both BaM and SrM (Figs.15 and 16), and neither contained the M-phase below 1100°C. However, the BaM or SrM phases were observed but hematite is the principal phase at all temperatures, representing as much as ~80% for BaM, or ~90 % for SrM.

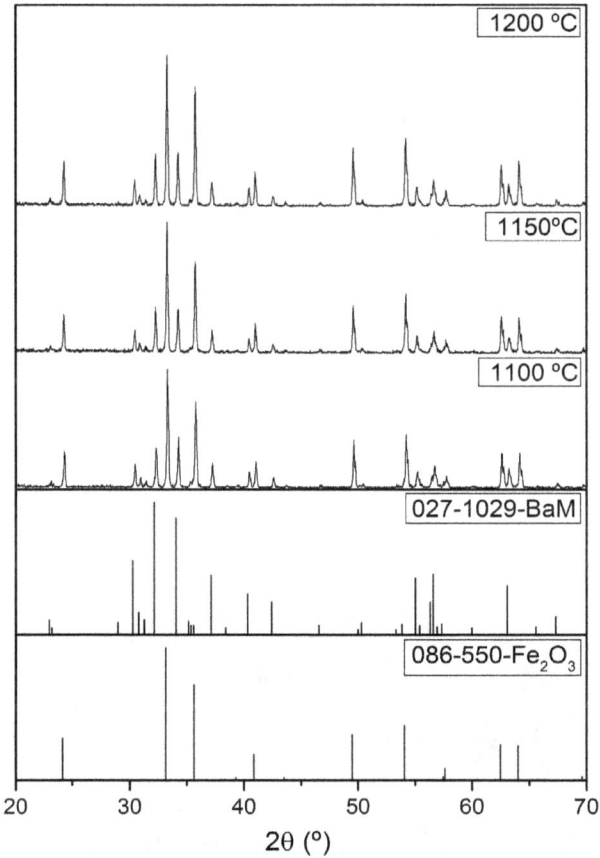

Fig.15. BaM hexaferrites prepared using sol-gel route and calcined at 1000°C, 1050°C, and 1200°C.

Table 1 summarises the phases present in the hexaferrites prepared from the various synthesis routes, with the major phases in bold. It can clearly be seen that the only route to produce a single phase M ferrite was the citrate combustion synthesis, and that coprecipitation and sol-gel routes gave a very-much mixed phase material. As long as it was at least 1100°C, the calcination temperature had relatively little effect. Therefore, ferrites calcined at 1200°C were used to prepare the composites.

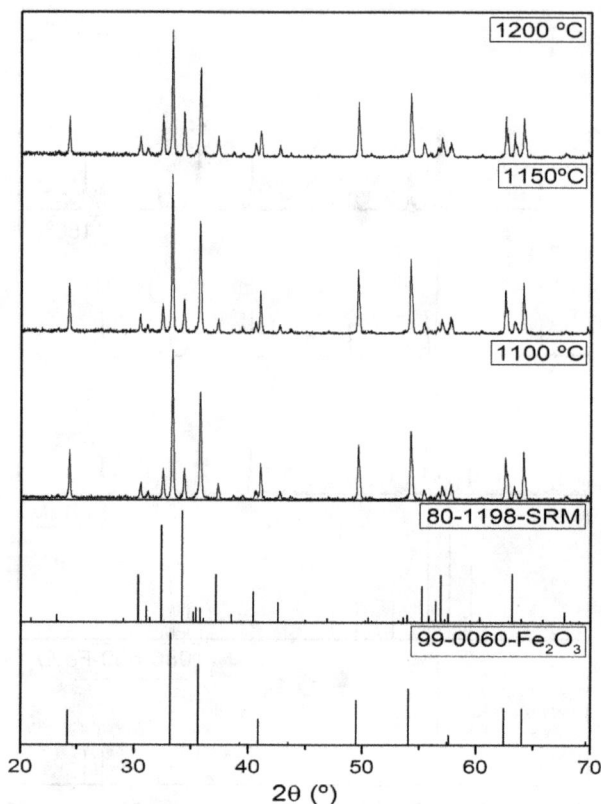

Fig.16. SrM hexaferrites prepared using sol-gel route and calcined at 1100°C, 1150°C, and 1200°C.

Table 1: Summary of the phases present in the hexaferrites prepared from different routes; major phases shown in bold.

Ferrite	Solid State	Coprecipitation	Citrate	Sol-gel
BaM	**BaM**, hematite	**Hematite**, BaM	**BaM**	**Hematite**, BaM
SrM	**SrM**, hematite	**Hematite**, SrM	**SrM**	**Hematite**, SrM

We made composites with ferrites from all four routes, as we were interested in the magentoelectric composites formed, and not just if the precursor ferrite powder was single phase or not.

4.2 XRD analysis of sintered barium/strontium hexaferrite-barium titanate composites

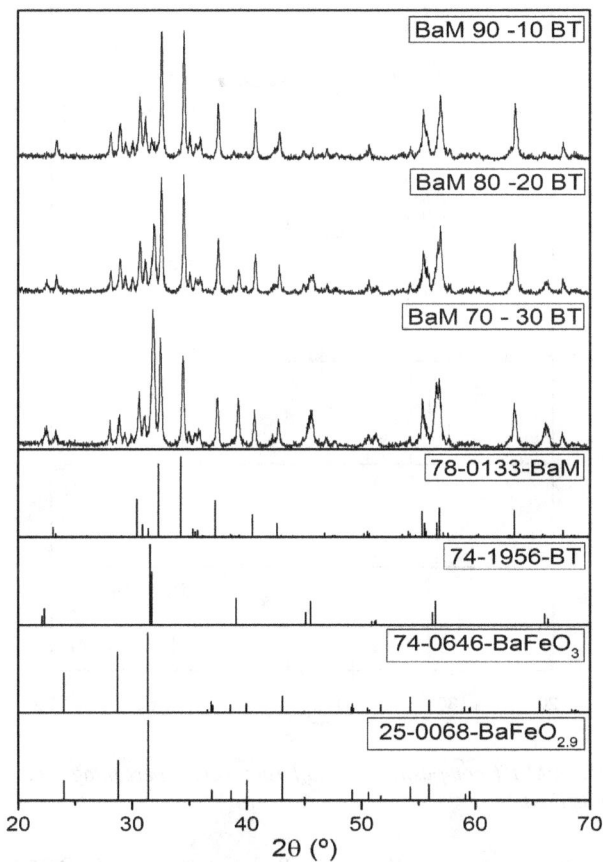

Fig.17. Sintered BaM-BT composites with different weight percentages, prepared by solid state reaction method.

XRD measurements were carried out of M-type hexaferrites and barium titanate composites to confirm the presence of both phases, in all three weight ratios used (90-10 wt %, 80-20 wt % and 70-30 wt % ferrite-BT), and the four different ferrite synthesis methods.

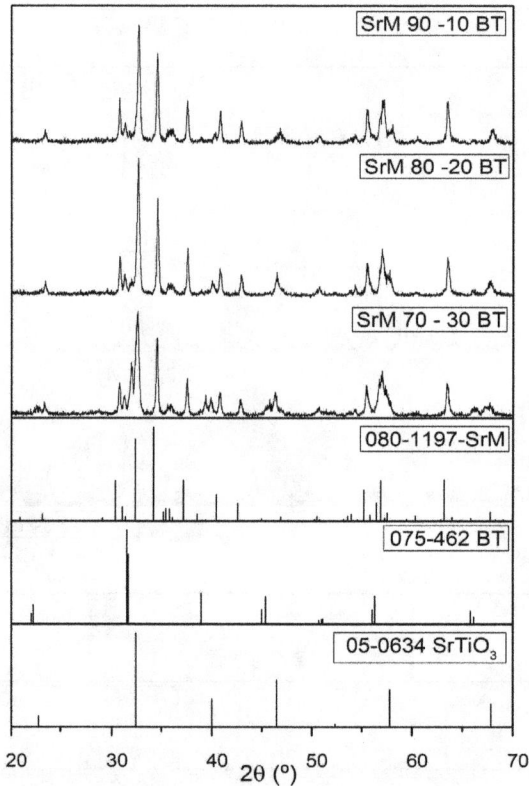

Fig.18. Sintered SrM-BT composites with different weight percentages prepared by solid state reaction method.

In the previous work reported by Karpinsky, Pullar *et al.* on 50-50 BaM-BT composites prepared by solid state reaction (using pure BaM powder), no evidence of reaction between the two phases was observed [23]. However, in these composites with greater amounts of BT, and therefore proportionately much greater quantities of barium in the composite, some other secondary phases were formed. This may also be due to the presence of hematite as a secondary phase in many of the ferrites used. With BaM made by solid state reaction, the BaM and BT are clearly the two major, and distinct, phases (Fig. 17).

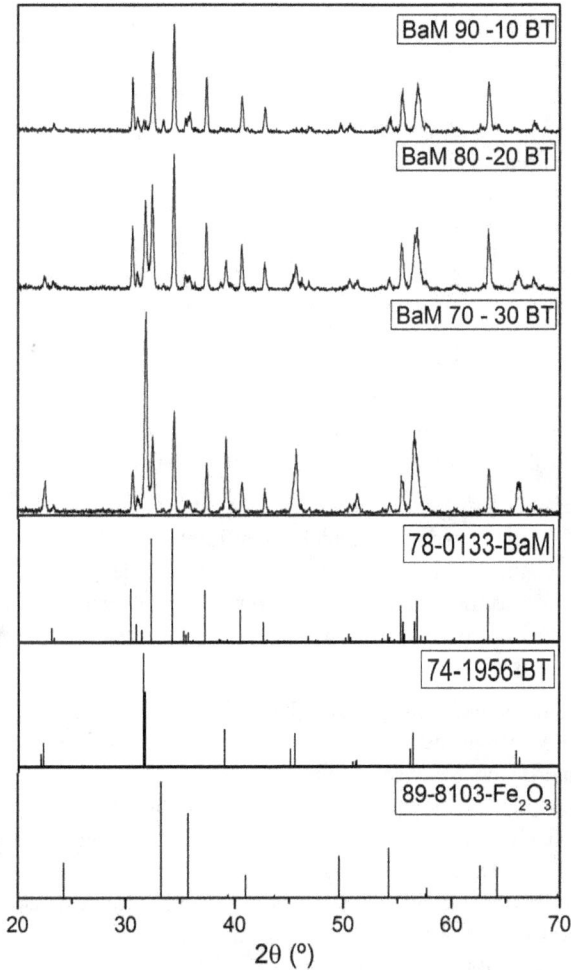

Fig.19. Sintered BaM-BT composites with different weight percentages, prepared by coprecipitation method.

However, diffusion of barium atoms has also led to the formation of a secondary phase, $BaFeO_3$, and possibly the oxygen-deficient $BaFeO_{2.9}$, with a small initial decrease in the quantity of the barium titanate phase. This indicates that BT is initially reacting with the hematite phase, which was present in the calcined ferrite powder, but is now absent from all samples. It would also appear that the BT does not then react with the BaM ferrite, as with increasing addition of BT, the 100% BT peak becomes significantly larger while the $BaFeO_3$ peak does not. This reinforces the previous findings that the BT and BaM phase do not, in fact, react with one-another.

While here have been several studies of BaM-BT composites, there has been almost no work on SrM-BT composites [24, 51]. In the composites of SrM-BT (Fig. 18), the formation of a secondary $SrTiO_3$ phase did occur.

The BT peaks were virtually absent in all samples, while the peaks for the $SrTiO_3$ phase increased with increasing BT addition. Despite the absence of any significant hematite peaks, no evidence was observed for the formation of either $BaFeO_3$ or $SrFeO_3$. This suggests that $BaTiO_3$ and SrM exchange atoms, resulting the formation of $SrTiO_3$, and the incorporation of Ba into the SrM lattice (a complete solid solution exists between BaM and SrM). Some released barium may even react with the hematite to form new BaM, explaining the absence of hematite, and the relatively low amounts of $SrTiO_3$ observed.

$SrTiO_3$ is a paraelectric material that exhibits significant piezoelectric characteristics only at very low temperatures, and it shows a very low piezoelectricity at room temperature. It is an incipient ferroelectric, approaching that state at around zero Kelvin. Therefore, it would not be expected to contribute greatly to the magnetoelectric properties or ME coupling in a composite.

Despite the M ferrites made from coprecipitation containing the least amount of ferrite phase, the sintered composites of BaM-BT consisted of nothing but the two distinct, discrete BaM and BT phases, with all three compositions tested (Fig. 19). The hematite peaks, which were the majority phase, are very small in the 90-10 composite, and almost disappear completely with increasing amounts of BT. This indicates that, far from reacting with BaM, the added BT actually has the effect of stimulating the further formation of the BaM phase in this case. This could be due to the much smaller particle sizes of a coprecipitated powder, even after calcination, encouraging diffusion and reaction to form the BaM ferrite in the presence of BT. No other barium ferrite phases were observed.

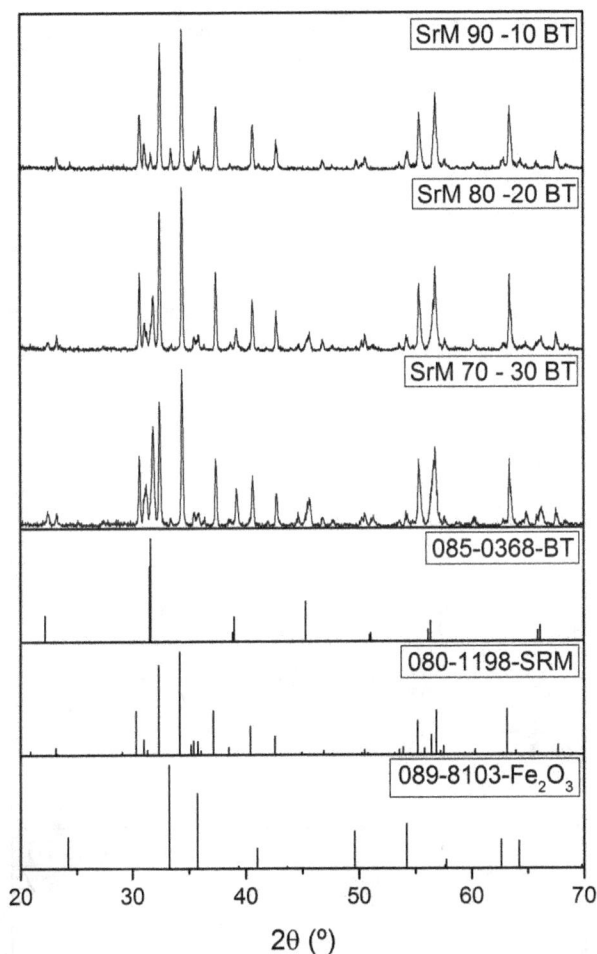

Fig.20. Sintered SrM-BT composites with different weight percentages, prepared by coprecipitation method.

A similar result was seen in the XRD patterns of the SrM-BT composites (Fig. 20). There is a presence of a very small amount of hematite as a secondary phase, which becomes smaller with increasing BT addition. The 100% BaTiO$_3$ peak at around 31° has displaced to the right by up to ~1°, indicating the probable diffusion of Sr^{2+} into the BaTiO$_3$ lattice to create $(Ba/Sr)TiO_3$ (BST). This BST is still a room temperature ferroelectric and piezoelectric phase. No evidence of $SrTiO_3$ was seen.

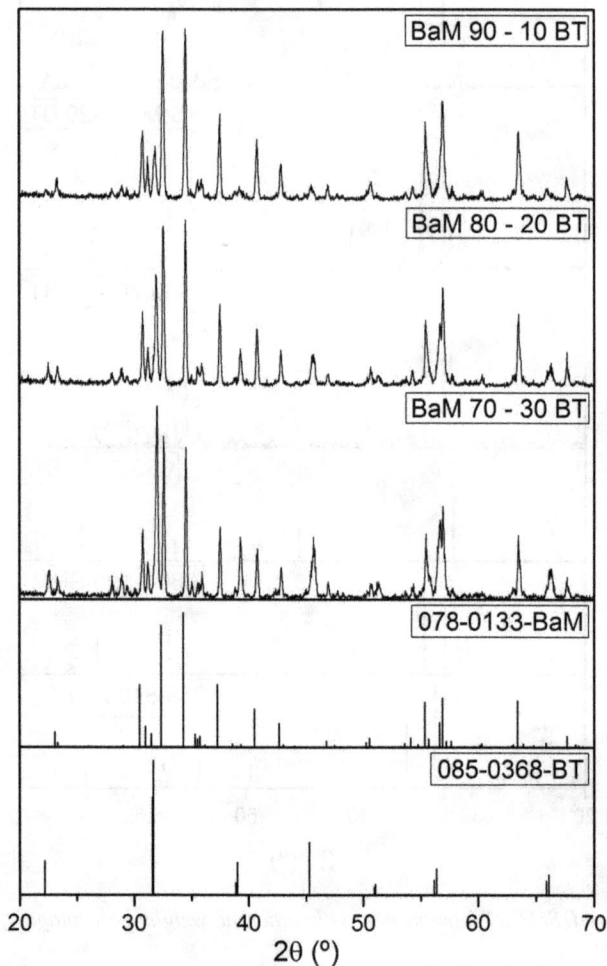

Fig. 21. Sintered BaM-BT composites with different weight percentages, prepared by citrates synthesis method.

The citrate route was the only one to result in single phase ferrites, and the XRD patterns of the BaM-BT composites from the citrates synthesis method show only two phases,

BaM and BaTiO$_3$ (Fig.21), the peaks of which are distinct, and no secondary phases are present (not even hematite), at all weight percentages. Therefore, the BT and BaM clearly do not react when present as single phases.

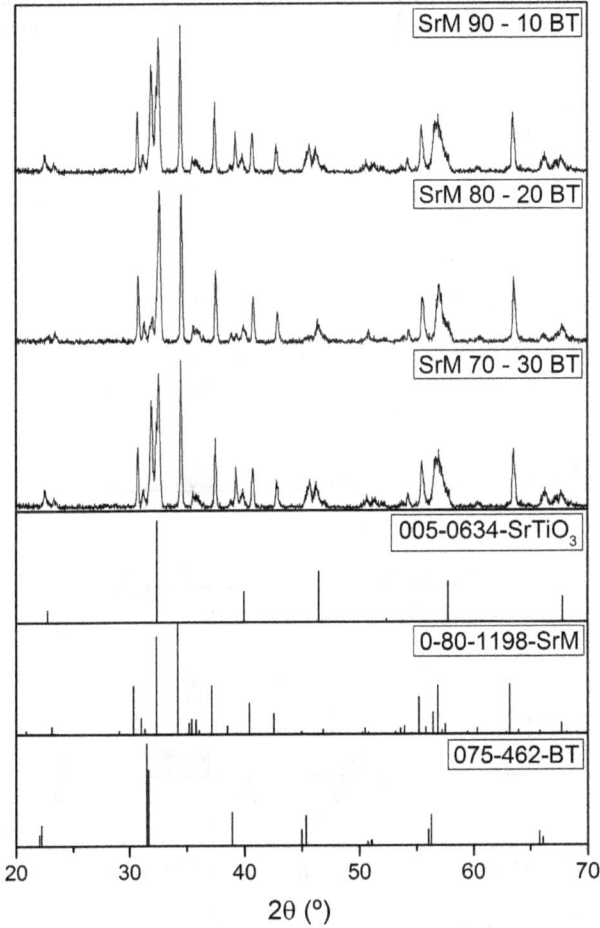

Fig. 22. Sintered SrM-BT composites with different weight percentages, prepared by citrates synthesis method.

However, in the XRD of SrM-BT composites, with SrM made from the citrate route, we can see that during the sintering a sizeable quantity of $SrTiO_3$ was formed, giving both $BaTiO_3$ and $SrTiO_3$ phases (Fig. 22). The peaks of this $SrTiO_3$ phase have a much greater intensity than those of $BaTiO_3$, making it very much the major non-magnetic phase. $SrTiO_3$ is a material that has very poor piezoelectric properties so we would expect these to be poor magnetoelectric composites.

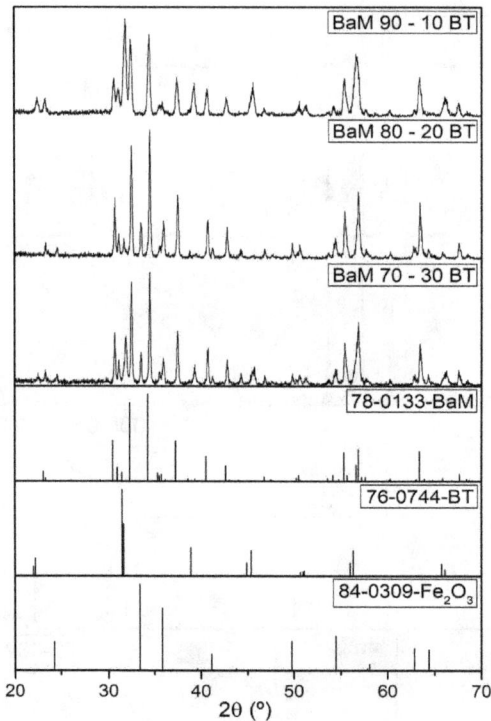

Fig. 23. Sintered BaM-BT composites with different weight percentages, prepared by sol-gel method.

Although there is still a small amount of $BaTiO_3$ phase present, the $SrTiO_3$ seems to form much more readily, suggesting that the formation of $SrTiO_3$ is greatly preferred to that of $BaTiO_3$ in these citrate-derived composites. The reason for this, compared to the coprecipitated SrM-BT composites, is unclear at present, but the major difference is the absence of any hematite phase in the ferrite precursor powder in this case.

In the sintering of the composites with BaM prepared by the sol-gel method, the patterns show BaM and BT as the major phases, with the minor presence of hematite in some cases (Fig. 23). Strangely, in the composite with the least amount of BT, the BT peaks are strongest, and hematite is absent. With increasing BT content, the intensity of the BT peaks reduced, while the hematite peak developed.

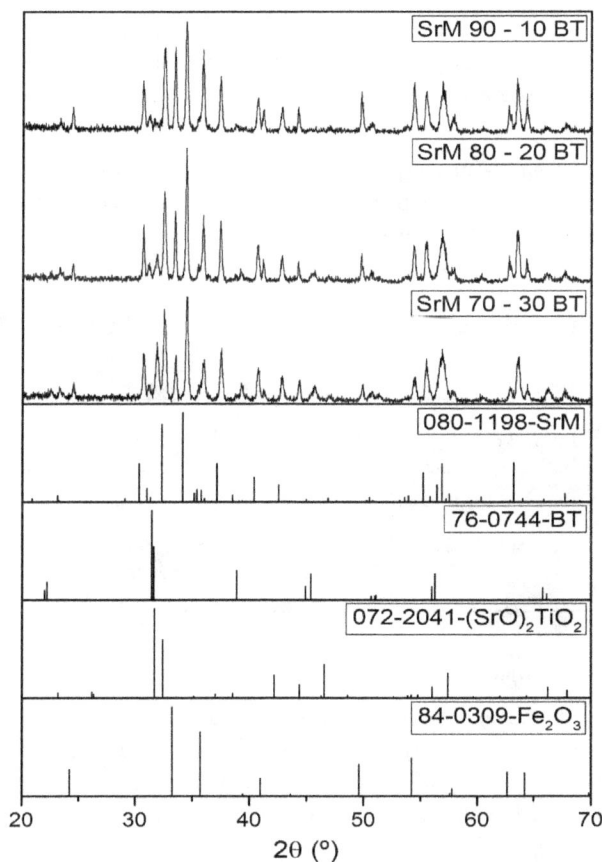

Fig. 24. Sintered SrM-BT composites with different weight percentages, prepared by sol-gel method.

35

The SrM-BT composites, made from sol-gel derived SrM, consisted of mostly SrM and hematite peaks, with more hematite present than that observed in the sol-gel synthesised SrM ferrite powder, as well as the BT phase (Fig. 24). As the quantity of BT increased, there was also the possible formation of a strontium titanate phase, but in this case not the perovskite $SrTiO_3$, but an unusual tetragonal (I4/mmm) $SrTiO_4$ structure. This is again quite different to the results for the other SrM-BT composites studied here, showing that the synthesis route for the SrM ferrite appears to have a major, and highly unpredictable, effect on the phases in the SrM-BT composite. It should be noted, however, that in these composites, if hematite was present, it was very much a minor phase, unlike in the ferrites prepared by the sol-gel route. Therefore, the composite mixture very much favours the crystallisation of the M ferrite phase at the expense of hematite in these samples.

In Table 2 is a summary of the phases present in composites of SrM and BaM with $BaTiO_3$, showing the principal phases in bold. The best methods for the different composites have the background in grey.

Table 2: Summary of the phases present in composites of BaM/SrM and BaTiO₃.

Composites	Solid state reaction	Coprecipitation	Sol-gel	Citrates
BaM + BT 70 – 30 80 – 20 90 – 10	**BaM, BT,** $BaFeO_{2.9}$, $BaFeO_3$	**BaM, BT**	**BaM, BT** Hematite	**BaM, BT**
SrM + BT 70 – 30 80 – 20 90 – 10	SrM, $SrTiO_3$, BT	**SrM, BT,** Hematite	**SrM, BT** Hematite	SrM, $SrTiO_3$, BT

4.3 Density of composites

To compare with the experimental values, the theoretical values of density were calculated for the composite mixtures, using the rule of mixtures, as shown in table 3. Published data for the density of the pure compounds from XRD pattern files and some bibliography was used. In this theoretical density we did not take into consideration any secondary phases which may have formed, just the primary ferrite and BT phases.

Table 3: Estimated theoretical density of the composites hexaferrite-barium titanate (BT) composites.

Theoretical composite density g cm^{-3}			
Mix	90 – 10	80 – 10	70 – 30
BaM	5.44	5.52	5.60
SrM	5.22	5.33	5.44

The density obtained using the uniaxial press (Table 4) was considerably lower than the theoretical density, being between about 56% to 75% of our estimated theoretical composite densities.

Table 4: Density of M-type hexaferrite - BT composites, uniaxial pressing.

Hexaferrites + BaTiO$_3$						
Uniaxial Press (density in g/cm^3)						
	Wt.%	Solid state reaction	Sol-gel	Coprecipitation	Citrates
BaM -BT	90 – 10	3.53	3.73	4.06	3.89
	80 – 20	3.62	3.99	4.09	3.99
	70 – 30	3.65	4.08	4.20	4.02
SrM -BT	90 – 10	3.05	3.77	4.00	3.53
	80 – 20	3.39	3.64	4.13	3.71
	70 – 30	3.52	3.89	4.09	3.69

With the density chart in Fig.25, is possible to visualise the difference of density between the ferrite synthesis methods used and the different composites. It can be seen that the coprecipitation route gave composites with consistently higher densities for both M ferrites, and that the solid state route gave the lowest densities. In general, the composites with BaM were slightly denser than their equivalents made with SrM, despite pure SrM reportedly having a lower sintering temperature by about 50-100°C [24]. However, the determining element for sintering of these composites is BT, which requires higher temperatures of around 1300°C and above to sinter as a pure ceramic [53]. The results here would seem to suggest that BaM can obtain a slightly denser composite with BaTiO$_3$ than SrM. Previous work by Karpinsky, Pullar *et al.* demonstrated that a 50-50 BaM-BT composite can be made from a standard solid state sintering route with about 85% density when sintered at 1250°C, although a minor amount of BaFe$_2$O$_4$ had already begun forming at this point, known to be a typical decomposition phase of BaM [23]. The maximum density expected in such co-sintered ceramic composites is about 90%. As a result, part of this work was to explore the synthesis of such composites at a lower temperature (1200°C), to avoid any such decomposition. The highest densities achieved with uniaxial pressing were with ferrites made by coprecipitation, with values of 4.20 g cm^{-3} = 75% for 70-30 BaM-BT, and 4.13 g cm^{-3} = 77 % for 80-20 SrM-BT.

With cold isotactic pressing, the densities of the composites in general increased (Table 5), compared with the results of the uniaxial press alone. Some of the composite densities increased by as much as 10 %, giving values of approximately 63% to 81% of the theoretical values.

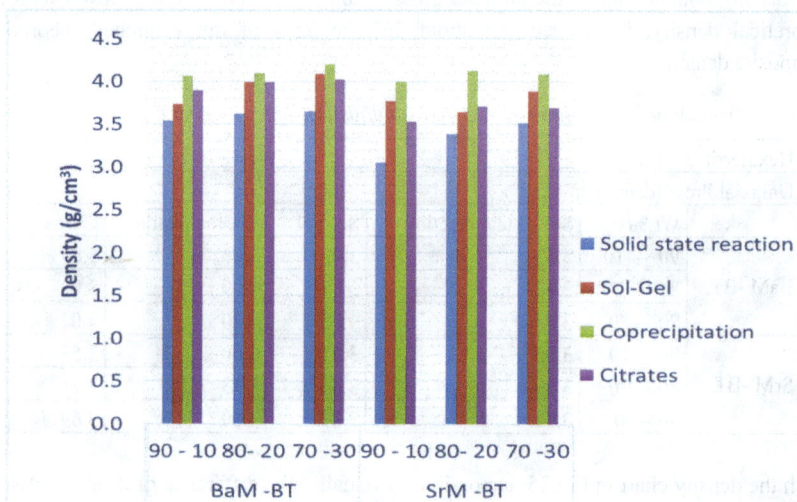

Fig.25. Density chart of M-ferrite - BT composites from uniaxial pressing.

The best densities achieved were in the 80-20 and 70-30 SrM-BT composites, with ferrites from the coprecipitation method, giving densities of 4.33 g cm^{-3} (81%) and 4.34 g cm^{-3} (80%), respectively. Similarly, the highest densities for the cold isotactic pressing of BaM-BT composites were in the 80-20 and 70-30 BaM-BT composites with coprecipitated ferrites, which both had densities of 4.31 g cm^{-3}, giving percentage densities of 78% and 77%, respectively.

Table 5: Density of M-type hexaferrite - BT composites through cold isotactic pressing

Hexaferrites + BaTiO$_3$					
Cold isotactic press (density in g/cm^3)					
	Wt.%	Solid state Reaction	Sol-gel	Coprecipitation	Citrates
	90 - 10	3.94	3.63	4.20	3.63
BaM -BT	80- 20	3.97	4.10	4.31	4.10
	70 -30	3.98	4.06	4.31	4.06
	90 - 10	3.56	3.80	4.22	3.80
SrM -BT	80- 20	3.83	3.96	4.33	3.96
	70 -30	3.97	4.03	4.34	4.03

In the chart in Fig. 26, it is possible to see that the densities are slightly higher, and that there is much less difference between the composites made with ferrites from different routes, or between the composites of different ratios. Nevertheless, it is clear that those made with coprecipitated ferrites still have the superior densities.

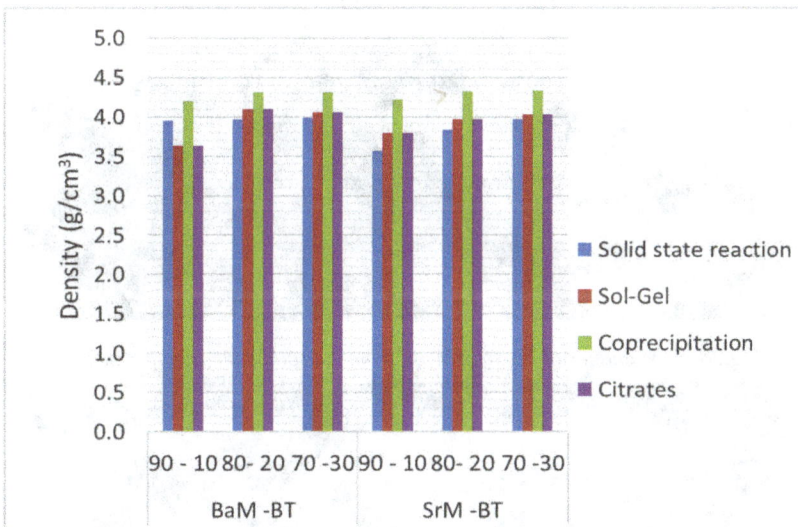

Fig. 26. Density chart of M-ferrite - BT composites from cold isotactic pressing

4.4 SEM images of m-ferrites-barium titanate composites

As the isostatic pressing resulted in slightly denser composites, the following SEM images (Fig. 27-31) were taken from various composites pressed in the cold isotactic press (CIP) before sintering.

Fig.27. SEM images of 70-30 BaM-BT composites from the solid state reaction method.

Fig.28. SEM images of 70-30 SrM-BT from the solid state reaction method.

Fig.29. SEM images of 70-30 BaM-BT composites from the sol-gel method.

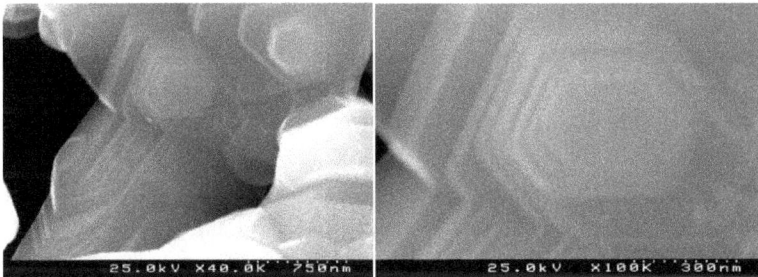

Fig.30. SEM images of BaM ferrite grains in the 70-30 BaM-BT composites from the sol-gel method.

Fig.31. SEM images of 70-30 SrM-BT composites from the sol-gel method.

With SEM images of M-hexaferrite-BT composites, it is possible to distinguish different phases by the shape of the grains. The hexaferrites have hexagonal unit cell symmetry, corresponding to the grains with hexagonal edges, and tend to form large grains, at least 0.5 µm diameter, with a plate-like form. Meanwhile, the grains with cubic symmetry represent the ferroelectric phase $BaTiO_3$, although this often appears as very small (hundreds of nanometres), irregularly shaped grains. The 70-30 BaM-BT composite from the solid state route (Fig. 27) presents a fairly typical microstructure, consisting of larger platy BaM grains measuring 0.5-2 μm in diameter, and smaller, much more equiaxed BT grains with diameters well below the micron level.

In the SEM images of the 70-30 SrM-BT composite from the solid state reaction (Fig. 28), it is possible to see that the hexaferrite grains are thicker and more elongated along one axis in the plane of the hexagon, compared to the BaM grains - presumably an influence from the effects of Ba substitution from the BT grains into the SrM matrix. All of the composites exhibit their high porosity / low density, and the grains do not appear to be orientated at all.

The BaM grains from the 70-30 BaM-BT composites made from sol-gel routes seemed to consist of much less plate-like ferrite grains, which were generally smaller with diameters around 0.5 μm and similar thicknesses (Fig. 29). However, when examined at higher magnification, it could be seen that these were in fact built of many repeated stacks of very thin hexagonal planes, these layers being tens of nanometres in thickness (Fig. 30). The SrM grains from the 70-30 SrM-BT composites made from the sol-gel route appeared even less like hexagonal plates, but were 0.5-1 μm, and the non-ferrite grains were still around 100 -300 nm (Fig. 31).

All the other composites had microstructures between these two extremes.

4.5   Vibrating sample magnetometry (VSM) measurements

VSM measurement provides information about the magnetic response of the composites, and with this technique it is possible to determine the saturation magnetisation (M_s), remnant magnetisation (M_r) and coercive field (H_c) from the magnetic hysteresis loop. Measurements of the 90-10 BaM-BT and SrM-BT composites were carried out at room temperature, with ferrites prepared using the sol-gel and solid state reaction routes. These were selected, as apart from the intrinsic properties of the magnetic phase, changes in microstructure can greatly affect the shape (coercivity) of the loop. Magnetic hysteresis loops are shown in Fig. 32, and the values of magnetic parameters are listed in Table 6.

As would be expected, all of the composites containing M ferrites were magnetically hard, with wide hysteresis loops. For pure M ferrites we expect an M_s of around 70 emu/g. Therefore, in a 90 % ferrite composite we would expect the maximum possible M_s to be 90% of this, i.e. about 63 emu/g. For the BaM composites, the solid state route gives a material with a reasonably high coercivity (283 mT), but a surprisingly low M_s. This is despite the fact that XRD indicate the iron-based phase to be single phase BaM. The loop from the sol-gel derived composite is even stranger, with a higher M_s but a lower coercivity of only 115 mT, verging on being a soft magnetic material. This larger magnetisation is strange, as XRD suggests that part of the Fe is in the form of hematite in this composite, but the lower coercivity could be explained by the much less plate-like form of these SrM grains formed of multiple layers of hexagonal planes.

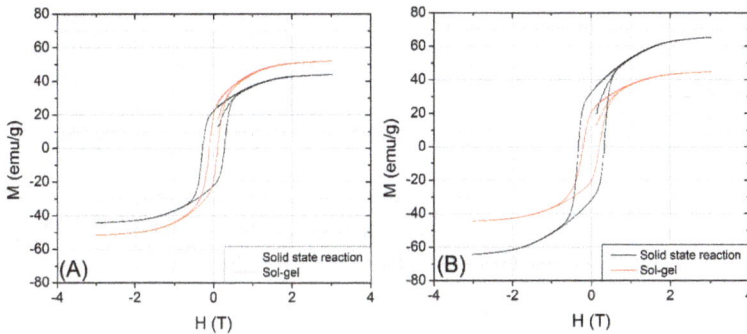

Fig.32. Magnetic hysteresis loops of 90-10 BaM-BT composites (A) and 90-10 SrM-BT composites (B), for the solid state reaction (black line) and sol-gel (red line) methods.

Table 6: Magnetic parameters - saturation magnetisation (M_s), coercive force (H_c) and remanent magnetisation (M_r) of M-type ferrite - BT composites, prepared using solid state reaction and sol-gel methods and pressed via cold isotactic pressing

Solid state reaction				Sol-gel			
Compositions 90 – 10 Wt.%	M_s (emu/g)	H_c (mT)	M_r (emu/g)	Compositions 90 – 10 Wt.%	M_s (emu/g)	H_c (mT)	M_r (emu/g)
BaM – BT	44	283	22	BaM - BT	51.8	115	18.7
SrM – BT	65.2	329	32	SrM - BT	44.6	210	20.3

For the SrM based composites, the opposite trend is seen, with that made from the solid state route possessing the highest magnetisation, near the maximum expected values, at 65.2 emu/g. Both this and the sol-gel route give better, harder magnetic loops, with coercivity values of 329 and 210 mT, respectively. Again, the less platy sol-gel composite gives a much lower coercivity than the plate-like SrM in the solid state reaction composite. The low M_s value of the SrM-BT sol-gel composite is due to the large amount of non-magnetic hematite it contains.

5. Conclusions

This article is concerned with the synthesis and characterisation of BaM/barium titanate and SrM/barium titanate composites. The motivation behind this work was the potential benefit of these composites for magnetoelectric applications. Structural analysis of composites based on hexaferrites prepared by four different synthesis routes were presented, and the magnetic properties of selected composites were discussed.

Acknowledgements

This work was developed within the scope of the project CICECO-Aveiro Institute of Materials, POCI-01-0145-FEDER-007679 (FCT Ref. UID /CTM /50011/2013), financed by national funds through the FCT/MEC and when appropriate co-financed by FEDER under the PT2020 Partnership Agreement. R.C. Pullar thanks the FCT for funding under grant IF/00681/2015. We are also extremely grateful for the infinite patience of Prof. R. B. Jotania.

References

[1] W.C. Röntgen, Ueber die durch Bewegung eines im homogenen electrischen Felde befindlichen Dielectricums hervorgerufene electrodynamische Kraft, Annalen der Physik, 271(10) (1888) 264-270. https://doi.org/10.1002/andp.18882711003

[2] P. Curie, Symétrie d'un champ électrique d'un champ magnétique,Journal de Physique Théorique et Appliquée, 3 (1894) 393-415. https://doi.org/10.1051/jphystap:018940030039300

[3] P. Debye, Bemerkung zu einigen neuen Versuchen über einen magneto-elektrischen Richteffekt, Zeitschrift für Physik A Hadrons and Nuclei, 36(4) (1926) 300-301. https://doi.org/10.1007/BF01557844

[4] A. Perrier and A.J. Staring, Archives des Sciences Physiques et Naturelles, 4 (1922) 373-382.

[5] E.O. Kamenetskii, M. Sigalov, and R. Shavit, Tellegen particles and magnetoelectric metamaterials, Journal of Applied Physics, 105(1) (2009) 013537-15. https://doi.org/10.1063/1.3054298

[6] V.J. Folen, G.T. Rado, and E.W. Stalder, Anisotropy of the magnetoelectric effect in Cr_2O_3, Physical Review Letters, 6(11) (1961) 607-608. https://doi.org/10.1103/PhysRevLett.6.607

[7] I.E. Dzyaloshinskii, On the magneto-electrical effect in antiferromagnets, Journal of Experimental and Theoretical Physics, 10(3) (1960) 628-629.

[8]   D.N. Astrov, The magnetoelectric effect in antiferromagnetics, Soviet Physics Journal of Experimental and Theoretical Physics, 11(3) (1960) 708-709.

[9]   G.A. Smolenskii and I.E. Chupis, Segnetomagnetics, Uspekhi Fizicheskikh Nauk, 137(3) (1982) 415-448. https://doi.org/10.3367/UFNr.0137.198207b.0415

[10]   Y.N. Venevtsev and V.V. Gagulin, Search, design and investigation of seignettomagnetic oxides,Ferroelectrics, 162(1)(1994) 23-31. https://doi.org/10.1080/00150199408245086

[11]   H. Schmid, Multiferroic magnetoelectrics, Ferroelectrics, 162(1)(1994) 317-338. https://doi.org/10.1080/00150199408245120

[12]   N.A. Hill, Why are there so few magnetic ferroelectrics?, The Journal of Physical Chemistry B, 104(29) (2000) 6694-6709. https://doi.org/10.1021/jp000114x

[13]   C.W. Nan, Magnetoelectric effect in composites of piezoelectric and piezomagnetic phases, Physical Review B, 50(9) (1994)6082-6088. https://doi.org/10.1103/PhysRevB.50.6082

[14]   C.W. Nan, M. I. Bichurin, S. Dong , D. Viehland and G. Srinivasan, Multiferroic magnetoelectric composites: Historical perspective, status, and future directions, Journal of Applied Physics, 103(3)(2008) 031101-36. https://doi.org/10.1063/1.2836410

[15]   J.V. Suchetelene, Product properties: a new application of composite materials, Eindhoven, 1972.

[16]   J. Boomgaard and R.A.J. Born, A sintered magnetoelectric composite material $BaTiO_3$-Ni(Co, Mn) Fe_2O_4, Journal of Materials Science, 13(7) (1978) 1538-1548. https://doi.org/10.1007/BF00553210

[17]   R.E. Newnham, D.P. Skinner and L.E. Cross, Connectivity and piezoelectric-pyroelectric composites, Materials Research Bulletin, 13(5) (1978) 525-536. https://doi.org/10.1016/0025-5408(78)90161-7

[18]   Jungho Ryu, Alfredo Vázquez Carazo[1], Kenji Uchino and Hyoun-Ee Kim, Magnetoelectric properties in piezoelectric and magnetostrictive laminate composites, Japanese Journal of Applied Physics,40 (2001) 4948. https://doi.org/10.1143/JJAP.40.4948

[19]   J. Van DenBoomgaard and R.A.J. Born, A sintered magnetoelectric composite material $BaTiO_3$-Ni(Co, Mn)Fe_2O_4, Journal of Materials Science, 13(7) (1978)1538-1548. https://doi.org/10.1007/BF00553210

[20]   J. Van Den Boomgaard, D.R. Terrell, R.A.J. Born, H.F.J.I. Giller, An in situ grown eutectic magnetoelectric composite material, Journal of Materials Science,9(10) (1974)1705-1709. https://doi.org/10.1007/BF00540770

[21]   R.C. Pullar and A.K. Bhattacharya, The synthesis and characterization of the hexagonal Z ferrite, $Sr_3Co_2Fe_{24}O_{41}$, from a sol-gel precursor. Materials Research Bulletin, 36 (7–8) (2001) 1531-1538. https://doi.org/10.1016/S0025-5408(01)00596-7

[22]   D. Karpinsky, E.K. Selezneva, I.K. Bdikin, F. Figueiras, K.E. Kamentsev, Y.K. Fetisov, R.C. Pullar, J. Krebbs, N.M. Alford and A.L. Kholkin, Development of novel multiferroic composites based on $BaTiO_3$ and hexagonal ferrites. Proceedings of the Materials Research Society, 1161 (2009) 7-12. https://doi.org/10.1557/PROC-1161-I01-06

[23]   D. Karpinsky, R.C. Pullar, Y.K. Fetisov, K.E. Kamentsev and A.L. Kholkin' Local probing of magnetoelectric coupling in multiferroic composites of $BaFe_{12}O_{19}$–$BaTiO_3$. Journal of Applied Physics, 108(4) (2010) 042012-5. https://doi.org/10.1063/1.3474967

[24]   R. C. Pullar, Hexagonal ferrites: a review of the synthesis, properties and applications of hexaferrite ceramics, Progress in Materials Science, 57 (2012) 1191-1334. https://doi.org/10.1016/j.pmatsci.2012.04.001

[25]   R.E. Newnham, Properties of Materials: Anisotropy, Symmetry, Structure, Oxford University Press, USA, 2005.

[26]   Robert C. Pullar, Pedro Marques, João Amaral and João A. Labrincha, Magnetic wood-based biomorphic $Sr_3Co_2Fe_{24}O_{41}$ Z-type hexaferrite ecoceramics made from cork templates, Materials and Design, 82 (2015) 297-303. https://doi.org/10.1016/j.matdes.2015.03.047

[27]   R.C. Pullar and R.M. Novais, Ecoceramics - cork-based biomimetic ceramic 3-DOM foams, cover article, Materials Today, 20 (2017) 45-46. https://doi.org/10.1016/j.mattod.2016.12.004

[28]   J. Smit and H.P.J. Wijn, Ferrites, Wiley, New York, 1959.

[29]   M. Pardavi-Horvath, Microwave applications of soft ferrites, Journal of Magnetism and Magnetic Materials, 215 (2000) 171-183. https://doi.org/10.1016/S0304-8853(00)00106-2

[30]   E.P. Wohlfarth and K.H.J. Buschow, Ferromagnetic materials: a handbook on the properties of magnetically ordered substances, North-Holland Publishing Co.1982.

[31] M. Sugimoto, Ferromagnetic materials, North Holland Physics
      Publishing,Amsterdam,1980.

[32] R. Gerber, Z. Šimša and L. Jenšovský, A note on the magnetoplumbite crystal
      structure, Czechoslovak Journal of Physics, 44 (1994) 937-940.
      https://doi.org/10.1007/BF01715487

[33] V. Adelsköld, X-ray studies on magneto-plumbite, $PbO.6Fe_2O_3$, and other
      substances resembling "beta-alumina", $Na_2O.11Al_2O_3$, Ark. Kemi Min. Geol.
      Series A-12. 29 (1938) 1-9.

[34] J.J. Went, G.W. Rathenau, E.W. Gorter and G.W. van Oosterhout, Hexagonal
      iron-oxide compounds as permanent-magnet materials.Physical Review, 86(3)
      (1952) 424-425. https://doi.org/10.1103/PhysRev.86.424.2

[35] R.A. McCurrie, Ferromagnetic materials: structure and properties 1994,
      Universidade de Michigan: Academic, 1994.

[36] International centre for diffraction data, USA, PDF no. 84-1531 ($SrFe_{12}O_{19}$), 84-
      757 ($BaFe_{12}O_{19}$), 84-2046 ($PbFe_{12}O_{19}$), Newton Square, PA.

[37] V.S. Puli, PTCR effect in La_2O_3 doped $BaTiO_3$ ceramic sensors, in Physics,
      Auckland University of Technology, New Zealand, 2006.

[38] A.D. Lozano-Gorrín, Structural characterization of new perovskites,
      polycrystalline materials - theoretical and practical aspects, InTech2012.

[39] M.M. Vijatovic, J.D. Bobic and B.D. Stojanovic, History and challenges of barium
      titanate: Part I, Science of Sintering, 40(2) (2008) 155-165.
      https://doi.org/10.2298/SOS0802155V

[40] M.S. Tadayuki Imai, Koichiro Nakamura, Kazuo Fujiura, Crystal growth and
      electro-optic properties of $KTa_{1-x}Nb_xO_3$, NTT Technical Review, 2007.

[41] L. Liliam Viana, Síntese e caracterização de compósitos titanato de bário-ferrita de
      cobalto preparados a partir de método sol-gel, Universidade Federal de Minas
      Gerais, 2010.

[42] S.M. Aygun, Processing Science of Barium Titanate, North Carolina State
      University, North Carolina,2009.

[43] R. Valenzuela, Magnetic Ceramics, Cambridge University Press,2005.

[44] R.C. Pullar, I.K. Bdikin, and A.K. Bhattacharya, Magnetic properties of randomly
      oriented BaM, SrM, Co_2Y, Co_2Z and Co_2W hexagonal ferrite fibres. Journal of the
      European Ceramic Society, 32(4) (2012) 905-913.
      https://doi.org/10.1016/j.jeurceramsoc.2011.10.047

[45] S. Sakka, Handbook of sol-gel science and technology, Characterization and properties of sol-gel materials and products, Kluwer Academic Publishers,2005.

[46] R.C. Pullar, M.D. Taylor and A.K. Bhattacharya, Novel aqueous sol-gel preparation and characterisation of barium M ferrite, $BaFe_{12}O_{19}$ fibres, Journal of Materials Science, 32 (1997) 349-352. https://doi.org/10.1023/A:1018593014378

[47] R.C. Pullar, S.G. Appleton and A.K. Bhattacharya, The manufacture, characterisation and microwave properties of aligned M ferrite fibres, Journal of Magnetism and Magnetic Materials, 186 (1998) 326-332. https://doi.org/10.1016/S0304-8853(98)00107-3

[48] R.C. Pullar, M.D. Taylor and A.K. Bhattacharya, A halide free route to the manufacture of microstructurally improved M ferrite ($BaFe_{12}O_{19}$ & $SrFe_{12}O_{19}$) fibres, Journal of theEuropean Ceramic Society, 22 (2002) 2039-2045.

[49] R.C. Pullar and A.K. Bhattacharya, Crystallisation of hexagonal M ferrites from a stoichiometric sol-gel precursor, without formation of the α-$BaFe_2O_4$ intermediate phase, Materials Letters, 57 (2002) 537-542. https://doi.org/10.1016/S0167-577X(02)00825-X

[50] M.P. Pechini, Method of preparing lead and alkaline earth titanates and niobates and coating method using the same to form a capacitor, U.S. Patent 3,330,697,(1967).

[51] Robert C. Pullar, Chapter 7 in: Mesoscopic Phenomena in Multifunctional Materials, A. Saxena and A. Planes (Eds), Springer, Heidelberg, 2014, pp. 159-200.

[52] Hsing-I Hsiang, Chi-Shiung His, Chun-Chi Huang, Shen-Li Fu,Sintering behavior and dielectric properties of $BaTiO_3$ ceramics with glass addition for internal capacitor of LTCC, Journal of Alloys and Compounds 459 (2008) 307-310. https://doi.org/10.1016/j.jallcom.2007.04.218

Chapter 2

Tuning the Magnetic Properties of M-type Hexaferrites

Sami H. Mahmood[1,a], Ibrahim Bsoul[2,b]

[1]Physics Department, The University of Jordan, Amman 11942, Jordan

[2]Physics Department, Al al-Bayt University, Mafraq 13040, Jordan

[a]s.mahmood@ju.edu.jo, [b]Ibrahimbsoul@yahoo.com

Abstract

In this chapter, common experimental techniques and preparation conditions adopted for the synthesis of M-type hexaferrites and their influence on the magnetic properties are briefly reviewed. The effects of various strategies of cationic substitutions on the properties of the hexaferrites are addressed. Further, our synthesis and findings on Co-Ti substituted hexaferrites are presented. It was found that Co-Ti substitution results in improving the saturation magnetization, and reducing the coercivity down to values favorable for high- density magnetic recording. Also, evidence of inter-particle interactions in the particulate samples was observed.

Keywords

Synthesis of Hexaferrites, M-type Hexaferrite, Structural Properties, Magnetic Properties

Contents

1. Introduction

Permanently magnetizable materials have acquired great scientific, as well as industrial and technological interest due to their crucial role in the fabrication of essential components for a multitude of devices and machines in active use nowadays. The search for new magnetic materials for high performance magnet applications has led to an exponential growth in both the scientific research in this field, and the investment in the development of such materials. This development in materials research was driven by the evolution of new technologies, and the ever increasing demand for the improvement in efficiency, better machine designs, and device miniaturization. The vigorous search for cost-effective magnetic materials had been driven by the large market share of permanent magnet industry, rated at more than \$15 billion in 2012, and expected to grow up to over \$28 billion by the year 2019 [1].

The discovery of M-type hexagonal ferrites ($MFe_{12}O_{19}$, where M = Ba, Sr, Pb) in the early 1950s can be considered one of the most important discoveries in the field of materials development, functionalization, and commercialization during the past few decades. The predominance of these materials in modern magnet industry since 1987 stems from their cost-effectiveness, in addition to their chemical stability, suitability for a wide range of applications, and high performance at relatively high temperatures [2-5]. The basic structural, electrical and magnetic properties of these ferrites are discussed in many previously published works [6-10].

The suitability of a magnetic material for specific applications is mainly determined by the requirements of these applications. For permanent magnet applications, high coercivity and high saturation magnetization are required. For magnetic recording purposes, however, the high coercivity requires an unfavorably high writing field, which entails high power consumption, and undesirable large device volume. The use of low coercivity material, on the other hand, may involve the risk of losing recorded data as a consequence of the susceptibility of the material to be demagnetized by stray fields. Consequently, a happy balance is called for. This requires modification and tuning the magnetic properties to fit the requirements of low power consumption and device miniaturization on the one hand, and minimize the risk of losing stored information on the other hand. The required value of the coercivity, however, depends on the specific application. While coercivities as low as 300 – 400 Oe could be used for low-density magnetic recording applications, higher coercivities in excess of 1000 Oe are required for high-density video recording applications [11]. The scenarios for tailoring hexaferrites with modified magnetic properties for specific applications included the adoption of different synthesis techniques to control the grain size and morphology of the produced ceramic, the variations of the stoichiometry of the starting powders and the experimental conditions, and the substitutions for the metal ions in the standard compound. These issues are addressed briefly in the following sections. Further, due to the scarcity of literature concerning the magnetic properties of Co-Ti substituted SrM hexaferrites, we dedicated forthcoming sections to the presentation and discussion of our findings concerning the structural and magnetic properties of these ferrites.

2. Synthesis techniques for the preparation of M-type hexaferrites

The conventional ceramic method is widely used for commercial production of hexaferrite powder. This technique, however, could be ineffective in controlling the grain size and morphology of the ferrite powder product. Accordingly, several other techniques involving chemical routes were adopted for the production of the magnet powders. The synthesis of ferrite powders by the various techniques often involved variations of the experimental conditions in determining the optimal conditions for the production of highly pure, high quality powder. Among others, the main factors taken into consideration in synthesizing the ferrite powders included: heat treatment, chemicals used in the starting powders, and stoichiometry of the precursor powder.

In the following subsections, the main techniques used to prepare ferrite precursor powders are briefly discussed. Further details on this matter are found in a previously published review article [2].

2.1 Conventional ceramic method

The standard ceramic method involving mixing and sintering appropriate molar ratios of metal oxides and carbonates precursor powders is a simple method which is widely used in the production of ferrite magnetic materials [12, 13]. Ball milling the starting powders provides smaller particle size and better homogeneity of the resulting ferrite precursor powder, allowing for better crystallization of the ferrite at lower sintering temperatures [2, 14]. Usually, the starting powders (usually α-Fe_2O_3 and divalent metal (Ba or Sr) carbonate) are mixed and dry-milled or wet-milled by balls in vials made of a mechanically hard material such as stainless steel, zirconia, and tungsten carbide. Wet milling in water, alcohol, or acetone media, for example, improves the milling efficiency. The ball-powder mass ratio is usually in the range 8 – 14, while the powder to liquid ratio is typically 1:1 [8, 15]. The product is then dried, and the resulting dry powder is compressed into desired shapes at high pressure for near full densification (~90% of the theoretical density of 5.28 g/cm^3 for BaM and 5.11 g/cm^3 for SrM). The compacts are then sintered at high temperature (900° C – 1300° C) for few hours. Compaction of wet-milled BaM powder at a pressure of ~ 250 MPa, and sintering at 1220° C for 3 h resulted in a density of the pellets of about 90% of the theoretical density [16], while compaction at ~ 500 MPa and sintering at 1100° C for 2 h was reported to result in ~ 80 – 93 % densification [17]. On the other hand, pellets compacted at 30 MPa and sintered at 1150° C were reported to have a density below 70% of the theoretical density [18].

The quality of the product was found to depend on the Fe:Ba molar ratio in the starting powder. Molar ratios of Fe:Ba $\geq$ 12 in the starting materials resulted in the coexistence of unreacted α-Fe_2O_3 phase with the major M-type phase [19-21]. In a detailed investigation of the effect Fe:Ba ratio on the quality of the synthesized powders, the optimal ratio for the production of a single BaM phase was found to be 11.7 [20]. On the other hand, a detailed and careful study indicated that Ba-rich starting powder mixtures (Fe:Ba < 11) resulted in the coexistence of $BaFe_2O_4$ spinel phase with the major M-type phase at temperatures < 1000° C, and the evolution of more complex oxides such as $Ba_3Fe_2O_6$ at higher temperatures [22].

2.2 Coprecipitation technique

This wet chemical technique allows for better reaction of the starting materials at the molecular level, and improves the product quality due to the homogeneity of the metal hydroxides co-precipitated in an aqueous solution of metal salts with the aid of a base (like NaOH) [23]. Appropriate molar ratios of metal chlorides or nitrates are usually dissolved to form homogeneous solutions which are subsequently mixed, and the base is then added drop by drop to promote the precipitation of metal powders. The co-precipitated powders are washed and dried, and the desired M-type hexaferrite is obtained by sintering this dried powder [24]. The required sintering temperature in this method is significantly lower than that required by the conventional solid-state reaction route [25, 26], and the starting solutions normally contain a Fe:Ba ratios lower than the theoretical stoichiometric ratio of 12 [27-30]. In a variation of the method [28], chloride gas was bubbled through a concentrated solution of NaOH to produce NaClO and NaCl solutions to which an appropriate amount of $Fe(NO_3)_3.9H_2O$ was added. Then $BaCl_2.2H_2O$ was added to the deep purple solution at Fe:Ba ratio of 10. The solution was left for 24 h, and then heated at 80° C for 1 h. The solutions were then filtered and rinsed to remove the residual chloride and alcohol. This process leads to the formation of crystalline barium hydroxide, and amorphous ferrihydrite which is structurally related to the hexagonal α-Fe_2O_3, and thus promotes the evolution of the BaM phase with heat treatment. The results of the study indicated the formation of a pure BaM phase at a low temperature of 800° C.

The effects of the experimental conditions on the quality and properties of the M-type powders prepared by this method were investigated by several workers [30, 31]. The increase in pH from 9 to 13, or the decrease of Fe:Sr ratio in SrM ferrites prepared by coprecipitation was reported to promote the formation of the M-type phase with reduced grain size and with the minimum coercivity of 4733 Oe and maximum saturation magnetization of 51 emu/g [30]. Further the grain size of BaM hexaferrite with Fe:Ba = 10 was reported to decrease with the increase of pH from 11 to 12.5, while for the samples with Fe:Ba = 10.5 or 11, the grain size increased with increasing pH [31]. The highest saturation magnetization of 66.1 emu/g was reported for the sample with Fe:Ba = 10 prepared at pH = 11 and sintered at 920° C, which decreased to 43.6 emu/g at pH = 12.5. On the other hand, increasing pH resulted in an increase in coercivity from 3400 Oe to 4334 Oe, which is consistent with the decrease of the particle size. The sample with Fe:Ba = 11, however, exhibited the highest coercivity of 4585 Oe at pH = 11, which decreased down to 4435 Oe with increasing pH value up to 12.5. Concurrently, the saturation magnetization of this sample decreased from 60.1 emu/g at pH = 11 to 46.2 emu/g at pH = 12.5.

In a yet another study, it was demonstrated that the saturation magnetization of the coprecipitated powder improved from about 25 emu/g to about 65 emu/g upon increasing the sintering temperature from 640° C to 920° C [32]. Also, the coercivity improved slightly from 5264 Oe at 640° C to 5791 Oe at 920° C, which is indicative of single domain particles with typical high coercivity. Drastically different results, however, were obtained for BaM with Fe:Ba = 12 at pH =12, and sintered at 1300° C, where a saturation magnetization of about 60 emu/g was found, but a rather low remanence of about 15 emu/g and coercivity of 860 Oe were reported [33]. The low coercivity is probably due to the large grain size of several microns in this sample.

2.3   Sol-gel method

This method is used to synthesize magnetic powders with controlled particle size distribution. In the standard technique, water solutions of metal chlorides or nitrates are mixed under constant stirring. In the citrate sol-gel method, an appropriate molar ratio of citric acid ($C_6H_8O_7$) is then added to the solution under constant stirring. The pH of the solution is adjusted in the range 7 – 9 by an addition of a basic solution drop wise with constant stirring [25, 34]. The solution is evaporated at about 80° C, and the resulting highly viscous gel is dried, and sintered to produce the required hexaferrite phase.

Modifications on the technique to improve the quality of the product, and investigate the effects of the experimental conditions were carried out by several investigators. In a variation on the technique, SrM ferrites were prepared by dissolving metal chlorides in citric acid to provide an acidic solution with very low pH, and the powders were subsequently calcined at 1000° C [35]. SrM ferrite prepared by this method exhibited a low saturation magnetization of 30.61 emu/g, and a coercivity of 2213 Oe. On the other hand, in a previous study [36], an aqueous solution of $Fe(NO_3)_3$ was precipitated with the aid of ammonia. The precipitate was then dissolved in citric acid with $BaCO_3$, keeping the molar ratio of Fe:Ba = 11.6, and ethylene glycol and benzoic acid were added to the transparent solution. The solution was evaporated at 60° C to produce a highly viscous gel, which was subsequently dried by heating at 170° C. The dried gel was then heat treated in two different ways. In *route a*, gel samples were placed in the furnace and the heating temperature was set at 1050° C with a heating rate of 4.5° C/min. Different samples were removed from the oven at different temperatures and quenched to room temperature in air. The samples obtained by this route were annealed at different temperatures for different periods of time. In *route b*, the samples were preheated at 450° C for 5 h prior to heating at temperatures in the range from 500° C to 1050° C for 5 h at the same heating rate. Samples prepared by *route a* in the temperature range of 300 – 500° C revealed the formation of γ-Fe_2O_3 and $BaCO_3$ phases and exhibited soft magnetic properties with specific magnetization (measured at 15 kOe) falling in the range 25 – 34 emu/g. The BaM phase developed at temperatures $\geq$ 550° C, and its fraction in the powder increased with increasing the heating temperature. In addition, α-Fe_2O_3

intermediate phase developed in the temperature range of 550 – 900° C, and $BaFe_2O_4$ was observed in the temperature range of 650 - 750° C. The coercivity increased sharply for powders annealed at temperatures above 650° C, reaching about 4200 Oe for powders annealed at temperatures 800 – 900° C. The saturation magnetization for powders heated at this temperature range also increased up to 65 emu/g, to be compared with 44 emu/g for the sample annealed at 650° C. At temperatures > 900° C, the coercivity decreased, reaching about 3400 Oe at 1050° C, and the specific magnetization increased up to ~ 69 emu/g. The samples prepared by route b exhibited a significant improvement of the magnetic properties in the whole temperature range, and optimum magnetic properties (specific magnetization of 70 emu/g and coercivity of 5950 Oe) were obtained in the range 900 - 950° C [36].

2.4   Auto-combustion method

In this modified citrate sol-gel method, self-propagating combustion of the gel occurs upon heating on a hot plate [37-39]. The starting homogeneous solution is prepared from metal nitrates and citric acid or ethylene glycol fuel with preset molar ratios, and the pH of the solution is adjusted in the range of 7 – 8 by adding a basic solution [40]. The solution is dehydrated by heating at 80° C resulting in a brown/yellow gel, which is subsequently heated at 220 – 240° C to promote auto-combustion yielding foamy powder [41, 42]. The ferrite powder is then obtained by subsequent grinding and sintering the powder at temperatures > 600° C.

Among the experimental variables which influence the properties of the prepared ferrite is the cation-to-fuel ratio and the Fe:Ba ratio [42]. An increase in this ratio from 1:1 to 1:2 was found to improve the magnetic properties, resulting in BaM with saturation magnetization of 55 emu/g, remanence of 28 emu/g, coercivity of 5000 Oe, and energy product of 1.013 MGOe, almost double the magnetic parameters obtained by using a 1:1 ratio [43]. Also, the properties of ferrite powders calcined at 900° C were found to be affected by the Fe:Ba ratio. Samples produced with Fe:Ba = 11 exhibited a saturation magnetization of 51 emu/g and coercivity of 4700 Oe, which improved to about 67 emu/g and 5650 Oe by using Fe:Ba = 9. In addition, the saturation magnetization increased with increasing the calcination temperature from 700° C to 1000° C, and the coercivity increased with increasing temperature up to 900° C, and then decreased down to 4500 Oe at 1000° C due to the presence of large particles growing at high temperatures.

2.5   Citrate precursor method

This method (also known as the Pecchini method) is used to prepare ultrafine barium ferrite powders at low temperatures via the decomposition of precipitated barium iron citrate complex [2, 44, 45]. The starting solution is prepared by dissolving stoichiometric ratios of Fe and Ba nitrates in deionized water, to which citric acid is added with the cation-to-citric acid molar ratio of 1:1. The solution is mixed while adding ammonia dropwise to increase the pH and improve the homogeneity of the solution. The solution is then heated at 80° C to improve the reaction and remove excess ammonia. Then barium iron citrate complex is precipitated by alcohol dehydration through the transfer of the solution drop by drop into ethanol with constant stirring. The yellowish precipitate is subsequently filtered and dried in an oven. The citrate precursor is decomposed at a temperature of 425 – 470° C, resulting in ~ 10 nm particles [44, 45]. These particles are normally superparamagnetic at room temperature, and subsequent sintering is required to obtain the BaM powder. Sintering the decomposed powder at 600° C resulted 50 nm particles with saturation magnetization of about 33 emu/g and a coercivity of about 580 Oe, whereas sintering at 700° C resulted in particle growth (to still below 100 nm), and a sharp increase in coercivity up to 4800 Oe, with a slight increase in saturation magnetization to 35 [44]. It is worth mentioning that the saturation magnetization is an intrinsic property of the material, while the coercivity depends on extrinsic parameters such as the particle size. Accordingly, the observed behavior of the magnetic parameters could be an indication that sintering at higher temperatures had the mere effect of increasing the particle size, transforming the powder from superparamagnetic in nature to an assembly of single domain particles with typically high coercivity. Further, it was demonstrated that BaM powder prepared by the citrate precursor method and sintering at 700° C was composed of 60 nm particles with the saturation magnetization of 61.5 emu/g, whereas samples fired at 750° C and 800° C exhibited particle growth into the range 80 – 100 nm [46].

2.6   Hydrothermal synthesis

In this method, aqueous solutions of metal nitrates are coprecipitated with a strong base solution such as NaOH or KOH with the appropriate $OH^-:NO_3^-$ and Fe:Ba ratios. The solution containing the metallic precipitates are then hydrothermally treated in an autoclave at temperatures in the range 150 – 290° C. The resulting particles are then filtered, washed and dried in an oven. To improve the magnetic characteristics of the product, the dried powder is sintered at temperatures of 1100 - 1200° C.

The experimental conditions including the hydroxide-nitrate ratio, the Fe:Ba ratio, the hydrothermal heat treatment temperature and duration, and the sintering temperature of the powder were found to be critical parameters for the properties of the final product [47]. Normally, hydroxide to nitrate ratio > 2 was found to be necessary for the formation of BaM, where lower ratios resulted in presence of intermediate iron oxide phase [47, 48]. However, hydrothermal treatment of aqueous solutions with much higher hydroxide-nitrate ratios (16) at temperature 150° C resulted in the formation of superparamagnetic particles ~ 10 nm in diameter [49, 50]. Further, the duration of the hydrothermal treatment was found critical for the formation of M-type phase. For example, samples prepared with Fe:Ba = 8 and OH^-:NO_3^- = 2 at 230° C were found to contain α-Fe_2O_3 as a major phase for heat treatment periods $\leq$ 10 h, whereas BaM phase was formed in the sample heat treated for 25 h [47].

The Fe:Ba ratio was also found to be crucial to the formation of the M-type using hydrothermal synthesis. A ratio < 10 was found necessary for the formation of BaM phase in samples prepared at a reaction temperature of 230° C for 48 h with fixed hydroxide-nitrate ratio of 2, where higher values resulted in the predominance of iron oxide in the product [47].

2.7   Molten salt method

This procedure is used to synthesize M-type hexaferrites with large crystals. The basic procedure involves mixing the reactants (barium carbonate and iron oxide precursor powders) with a solvent consisting of NaCl-KCl salt mixture, and heating the reaction mixture at 800 - 1100° C [51]. The procedure results in a dry cake in which the magnetic hexaferrite powder is entrapped. The dried body is then crushed and washed with distilled water to remove the salt, and the magnetic powder is obtained. Large variations of the magnetic properties were revealed depending on experimental conditions including the starting reactants, solvent composition and purity, and the heat treatment [51, 52]. Hexagonal plates of BaM with basal dimension < 1.5 μm and optimal magnetic properties (saturation magnetization of 72 emu/g, and H_c = 4300 Oe) were synthesized under optimal experimental conditions [51].

In a variation to the method, BaM powder is first prepared by coprecipitation, and the coprecipitated particles are mixed with KCl flux at a BaM to salt weight ratio of 1:1 [53]. BaM particles were synthesized by initially heating the mixture at 450° C, and then at 950° C. The product was then washed with deionized water to remove the salts, and dried at 80° C in an oven.

2.8   Glass crystallization method

In this method, the ferrite powder is mixed and melted with a glass matrix. Subsequently, the melt is rapidly quenched to produce an amorphous matrix containing the ferrite. The crystallization of the hexaferrite phase is achieved by annealing at temperatures > 600° C, and the magnetic ferrite is retrieved by dissolving the amorphous matrix in a dilute acid, which does not dissolve the hexaferrite phase [54, 55]. Samples prepared by this method at different heat treatments revealed coercivity ranging from 2600 Oe to 5350 Oe [56].

2.9 Spray pyrolysis method

A system consisting of an ultrasonic droplet generator, a quartz high-temperature reactor, and a powder collector was used for the preparation of hexaferrite [57-59]. The droplets of the precursor solution are injected into the reactor by a gas flow, which can be adjusted for optimal results. The evaporation of the droplets at 900° C resulted in a powder consisting of spherical particles. Improvement of the magnetic properties was achieved by subsequent heat treatment. As expected, the post heat treatment influenced the particle size and coercivity greatly, and the sample heat-treated at 800° C revealed a high coercivity of 6000 Oe [60].

2.10 Further developments in synthesis routes

The pursuit of high-quality hexaferrite materials with high magnetic properties have led to the development of a variety of powder synthesis routes, some of which were discussed in the previous subsections. The final product, depending on the stoichiometry of the starting powder or solution mixtures and the adopted experimental conditions, often resulted in secondary phases which influenced the magnetic properties negatively. Accordingly, various remedies were proposed to improve the purity of the ferrite powder and enhance its magnetic properties. These include modifications of the prevailing experimental conditions and combining synthesis routes, as well as proposing new synthesis methods.

Barium hexaferrite powder prepared by the conventional ceramic method and calcined at 900° C was reported to be composed of a mixture of equilibrium phases, consisting of BaM and the intermediate $BaFe_2O_4$ and Fe_2O_3 phases [61]. Small additions of B_2O_3, however, was found to result in a single BaM phase with enhanced remanent magnetization at such low sintering temperature. Specifically, the sample with 1 wt.% B_2O_3 addition revealed remanence magnetization of 28 emu/g, saturation magnetization of 54 emu/g, and coercivity between 2000 and 3000 Oe; these properties are suitable for magnetic recording applications. Further improvements of the magnetic properties were achieved with B_2O_3 addition and etching with diluted HCl solution [62]. Specifically, the sample with 0.1 wt. % B_2O_3 exhibited a rather high remanent magnetization of 34.9 emu/g and magnetization of 63.3 emu/g measured at 1.5 T applied field. The coercivity, however, did not exhibit a systematic behavior with HCL washing, and the enhancement of the remanence and saturation magnetization were attributed to the removal of nonmagnetic impurity phases which may not be detectable by XRD.

The coercivity of BaM magnets prepared by ball milling and sintering was also found to be enhanced by small additions of V_2O_3 due to the formation of finer powders with smaller average particle size and narrower particle size distribution [63]. At low V_2O_3 concentrations (0.7 wt.%) BaM samples sintered at 1100° C reveled an increase of the coercivity from 3.5 kOe to 4.1 kOe, while the saturation magnetization and remanent magnetization remained almost the same as for the un-doped sample (68.6 emu/g and 37.4 emu/g, respectively). At 3.5 wt.% V_2O_3, however, significant fractions of nonmagnetic secondary phases were observed, and the saturation magnetization and remanent magnetization reduced dramatically down to 21.8 emu/g and 11.4 emu/g, respectively. Higher sintering temperatures improved the saturation magnetization (26.6 emu/g at 1300° C) and reduced the coercivity down to 1.6 kOe. The saturation magnetization and remanence were improved significantly (to 59.6 emu/g and 32.8 emu/g, respectively) by adding an extra 14 wt.% $BaCO_3$ (for (Fe+V):Ba = 6.3), sintering at 1200° C, and washing with diluted HCl solution. The coercivity of the product was 2.1 kOe, and the relatively high remanent magnetization makes the product suitable for magnetic recording media.

The oxalate precursor route adopted for the synthesis of BaM ferrite was found to produce a powder with a rather low coercivity [64]. In this method, metal chlorides with Fe:Ba = 12 were dissolved in equal amounts of oxalic acid, and mixed with continuous stirring using a magnetic stirrer for 15 min. The solution was then heated at 80° C with constant stirring and dried at 100° C overnight. The dried powder precursors were then annealed at temperatures in the range of 800 – 1200° C for 2 h. The prepared powders exhibited an increase of both the saturation magnetization and coercivity with increasing the annealing temperature from 900 to 1100° C. The maximum saturation magnetization was found to be 66.36 emu/g, while the maximum coercivity was only about 640 Oe.

The direct crystallization of BaM ferrite from an aerosol was realized for the first time by a combination of pyrolysis and citrate precursor methods [65]. The barium iron citrate precursor solution with Fe:Ba = 12 was nebulized at a flow rate of 0.5 ml/min to produce a stream of fine droplets, which was passed through the low-temperature furnace (200 – 250° C) of the reactor to evaporate the solvent. The dried barium iron citrate particles were subsequently passed through a high-temperature furnace, resulting in a powder consisting of submicron hollow spheres with a rather low saturation magnetization of 5.6 emu/g and a coercivity of 2500 Oe. Further heat treatment of the powder at 1000° C improved the magnetic properties significantly, where the saturation magnetization increased up to 50.0 emu/g and the coercivity to 5600 Oe. When the powders were hand milled to eliminate the effects of particle aggregation, the coercivity of the heat-treated sample further increased to 5900 Oe, but the saturation decreased to 42.6 emu/g. The obtained coercivity by this synthesis route is probably one of the highest reported for pure BaM ferrite.

In addition, a new synthesis route, named ammonium nitrate melt technique (ANMT) [66], was proposed for the production of high-quality BaM ferrite using low Fe:Ba ratios [67]. In this method, the desired proportions of $BaCO_3$ and Fe_2O_3 powders were mixed with the ammonium nitrate melt and stirred with a magnetic stirrer. The resulting thick solution was evaporated at 260° C, resulting in a reddish precipitate which was subsequently preheated at 450° C for 5 h. Parts of the resulting powder were then heat treated at temperatures in the range of 800 – 1200° C to identify the optimal heat treatment. The results of the study revealed that the powder with Fe:Ba ratio of 2 gives the highest saturation magnetization. When this powder was sintered at 1100° C, a mixture of $BaFe_2O_4$ and BaM phases was observed. Upon washing the powder with HCl solution, the $BaFe_2O_4$ phase disappeared, and a single BaM phase with the saturation magnetization of 66.7 emu/g, remanent magnetization of 38.5 emu/g and coercivity of 4228 Oe was obtained.

The above discussion indicates that the synthesis method and the experimental conditions play a critical role in the quality and magnetic properties of the produced hexaferrite powder. Hexaferrites with high coercivity and saturation magnetization are suitable for a wide variety of permanent magnet (PM) applications, including low-power motors in home appliances, motors and switches in auto-industry, loud speakers, current meters, magnetic seals, magnetic shielding, toys, etc. Ferrites with lower coercivities, however, are suitable for magnetic recording applications. While materials with coercivities in the range of few hundred Oe can be used for low density longitudinal magnetic recording (LLMR) media, materials with higher coercivities up to 1200 Oe can be used for high-density longitudinal recording media (HLMR). Materials with higher coercivity are not suitable for longitudinal recording, and can effectively be used for high-density perpendicular recording (HPMR) applications. The production of a hexaferrite material with the desired properties requires optimization of the experimental procedures to produce high quality single-domain powders with controlled particle size. In Table 1, some of the experimental findings concerning the magnetic properties of M-type hexaferrites with the potential permanent magnet and magnetic recording applications are presented.

Table 1: Magnetic properties of M-type hexaferrites prepared by different techniques.

Synthesis Method	M_s (emu/g)	M_r (emu/g)	H_c (Oe)	Application	Reference
Conventional Solid State Method	61	32	2080	HPMR	[68]
	49	24	1005	HLMR	[69]
	57		1943	HPMR	[70]
	72		4200	PM	[71]
	71	37	4020	PM	[72]
	60		4000	PM	[73]
Coprecipitation	64	31	4700	PM	[74]
	72		5340	PM	[75]
	69	36	5440	PM	[76]
	65		5540	PM	[77]
	70		5044	PM	[25]
	60	15	860	HLMR	[33]
	71		6400	PM	[78]
	65	36	5791	PM	[32]
Sol-gel	61		1827	HPMR	[79]
	49		4800	PM	[34]
	61	37	4996	PM	[80]
	59	36	1920	HPMR	[81]
	61	36	5692	PM	[82]
	70		5900	PM	[83]
	70		5950	PM	[36]
	67		5650	PM	[42]
Auto combustion	55	28	5000	PM	[43]
	51	36	2037	HPMR	[40]
	64		750	HLMR	[84]
	50	31	5017	PM	[85]
	60		4250	PM	[86]
	40	22	5689	PM	[87]
	74		5163	PM	[88]
Hydrothermal	40		2500	PM, HPMR	[49]
	59	20	1350	HPMR	[89]
	64		2300	PM, HPMR	[47]
Molten salt	59		4820	PM	[90]
	72		4650	PM	[53]
	72		4300	PM	[51]

3. Effects of cationic substitution

In the preceding section, we addressed the effects of synthesis method and experimental conditions on the magnetic properties of hexaferrites. It was demonstrated that great variations of the magnetic properties were achieved through modifications of the experimental procedures. However, such variations were mainly due to the quality and purity of the produced hexaferrite, and the particle size and morphology of the powder. An efficient scenario for the modification of the magnetic properties can also be achieved by cationic substitution in the standard M-type hexaferrite, which normally results in modifying the intrinsic properties such as the magnetocrystalline anisotropy [91, 92].

Cationic substitution in M-type hexaferrite could involve the divalent (M^{2+}) ions, and/or the Fe^{3+} ions. Normally, Sr is used to replace Ba in BaM-type hexaferrite [59, 93, 94], where SrM hexaferrite with an improved coercivity of $\geq$ 6.4 kOe was prepared by different experimental procedures [95, 96]. The use of Pb^{2+} and Ca^{2+}, on the other hand, generally result in lowering the coercivity of M-type hexaferrite [97-99]. In an earlier publication, however, the addition of CaO was reported to improve the coercivity of SrM ferrite [100]. Further, the effect of Ca and Ca-Sr substitution for Ba in BaM ferrite prepared by the conventional ceramic method, pre-sintering at 600° C, and sintering at 1100° C was investigated [101]. The results of the study revealed an increase in coercivity from 2.75 kOe for BaM to 3.2 kOe for $Ba_{0.5}Ca_{0.5}Fe_{12}O_{19}$, with a concurrent decrease in saturation magnetization from 53.04 emu/g to 33.17 emu/g. The sample $Ba_{0.5}Ca_{0.25}Sr_{0.25}Fe_{12}O_{19}$, on the other hand, exhibited a higher saturation magnetization of 50.20 emu/g and a lower coercivity of 2.98 kOe.

The effects of the substitution of the divalent metal ions by rare-earth ions were also given some attention. The substitution of Ba by Eu in BaM prepared sol–gel method was found to result in an increase of the coercivity from 1.92 kOe for the un-doped sample up to 6.12 kOe for $Ba_{0.75}Eu_{0.25}Fe_{12}O_{19}$ [81]. The saturation magnetization at this stoichiometry dropped down from 59.21 emu/g to 45.14 emu/g, while the remanent magnetization decreased slightly from 35.66 emu/g to 34.52 emu/g. Also, it was found that only 10% substitution of Ba by La in BaM prepared by a reverse micro-emulsion route leads to high magnetic properties of M_s = 66.5 emu/g, M_r = 34.6 emu/g, and H_c = 5229 Oe [102]. In addition, BaM nanofibers prepared by citrate sol–gel method and electrospinning, followed by heat treatment, revealed an increase in saturation magnetization from about 64.5 emu/g up to 77.2 emu/g and a decrease in coercivity from 4323 Oe down to 3565 Oe at 5% La substitution for Ba [103]. At 10% La substitution, the saturation magnetization dropped down to 71.0 emu/g, and the coercivity increased to about 4160 Oe. The results of this study also revealed an enhancement of the microwave absorption properties with La substitution for Ba, and the La-doped compounds were reported to have a significant potential for microwave absorption applications. On the other hand, the substitution of Sr by La–Ce combination in Zn-substituted SrM ferrite prepared by the conventional solid state route was reported to induce a reduction in both the saturation magnetization and coercivity of the hexaferrites calcined in air at 1200° C [104]. The saturation magnetization of the calcined powders was enhanced by annealing at 1100° C in the N_2 environment, and a product with saturation magnetization as high as 77.4 emu/g and coercivity of 2343 Oe was obtained for $Sr_{0.7}La_{0.1}Ce_{0.2}Fe_{11.7}Zn_{0.3}O_{19}$ ferrite. Also, in a recent publication, it was demonstrated that the saturation magnetization improved upon increasing x up to 0.2, and the coercivity increased with increasing x up to 0.3 in $Sr_{1-x}La_xFe_{12-x}Co_xO_{19}$ [105]. Further, partial substitution of Ba by either La or Pr in BaM powders prepared by auto-combustion was reported to result in an improvement of the saturation magnetization and coercivity of the ferrite, Pr substitution being more effective, especially in increasing the coercivity [106].

The substitution of Fe^{3+} by a trivalent metal, or by combinations of divalent and tetravalent metals was extensively investigated for the purposes of modifying the magnetic properties to fit specific applications. The effects of such substitutions will be the subject of the following subsections.

3.1 Fe^{3+} substitution by trivalent metal ions

In general, the substitution of Fe^{3+} by a trivalent metal such as Al, Cr, or Ga results in an increase in coercivity, with Al being the most effective [68, 71, 107-109]. Since the coercivity is influenced by the particle size, a reduction in coercivity is normally observed at elevated sintering temperatures due to particle growth beyond the critical single domain size [110]. Since the early times of BaM synthesis and characterization, small additions of kaolin ($Al_2O_3(SiO_2).2H_2O$), however, were reported to inhibit particle growth with increasing the sintering temperature and improve the magnetic properties [13]. The improvement in the maximum energy product $(BH)_{max}$ with kaolin addition, however, was not significant, where it ranged from 3.34 MGOe with 1 wt.% kaolin to 2.55 MGOe with 3 wt.% kaolin addition; this is to be compared with values in the range of 2.97-3.52 MGOe for BaM with no kaolin addition and sintering at different temperatures [13]. This is possibly due to the competing effects of increasing the coercivity and decreasing the saturation magnetization with kaolin addition.

$BaFe_{10}Al_2O_{19}$ prepared by ball milling and sintering at 1100° C revealed an increase in coercivity up to 9.3 kOe, accompanied by a significant reduction of the saturation magnetization down to 36 emu/g [108]. Also, a coercivity of 7.1 kOe and a saturation magnetization of 21.6 emu/g were reported for $SrFe_{10}Al_2O_{19}$ ferrite prepared by auto-combustion and calcination at 950° C, and a record coercivity of 16.2 kOe with a rather low saturation magnetization of 11.80 emu/g were reported for $SrFe_8Al_4O_{19}$ [111]. In addition, an increase in coercivity up to 7.4 kOe, and a reduction in saturation magnetization down to 36.50 emu/g were observed in $SrFe_{10}Al_2O_{19}$ prepared by auto-combustion and calcination at 1100° C [112]. Further, $SrFe_{10.7}Al_{1.3}O_{19}$ prepared by glass crystallization and calcination at 950° C revealed a high coercivity of 10.18 kOe, accompanied by a reduction in saturation magnetization down to 18.4 emu/g [113]. Furthermore, a relatively high saturation magnetization of 50.5 emu/g with the coercivity of 6.0 kOe was obtained in $BaFe_{11.2}Al_{0.8}O_{19}$ prepared by ball milling and calcination at 1100° C [71]. However, contradictory results on Al-doped BaM were reported elsewhere [114].

The coercivity of Al-substituted SrM ferrites was improved by the partial substitution of Sr by a rare-earth (RE) element. A systematic study of the effect of the type of RE ion substitution on the magnetic properties of $Sr_{0.9}RE_{0.1}Fe_{10}Al_2O_{19}$ revealed a significant improvement of the coercivity of SrM ferrite, Pr being the most effective in enhancing the coercivity from 7.4 kO up to 11.0 kOe [112]. The saturation magnetization, on the other hand, decreased from 36.5 emu/g down to 30.8 emu/g with Pr substitution. At this point, it is worth mentioning that although Al^{3+} substitution resulted in a significant increase in coercivity, the effect of such substitution leads to deterioration of the saturation magnetization, and limits the applicability of the product to applications demanding high coercivity, without necessarily very high saturation magnetization values. Such applications include the production of permanent magnets for devices operating for long times in environments of high stray fields.

Cr^{3+} substitution for Fe^{3+} was also reported to enhance the coercivity of M-type hexaferrite [115, 116]. Cr-substitution for Fe in BaM ferrite ($BaFe_{12-x}Cr_xO_{19}$) prepared by sol–gel auto-combustion method and calcination at 1100° C revealed a decrease in saturation magnetization and an increase in coercivity with increasing Cr content up to x = 0.8 [116]. The decrease in saturation magnetization was attributed to magnetic dilution or spin canting as a result of Cr substitution for Fe at 2a and 12k sites of the hexaferrite lattice. The increase in coercivity from 2.0 kOe at $x = 0$ up to 5.2 kOe at $x = 0.8$, on the other hand, was associated with the reduction in grain size with increasing Cr content. Similar effects on grain size and morphology, and on the magnetic properties of Cr substitution for Fe in SrM ferrite ($SrFe_{12-x}Cr_xO_{19}$) prepared by microwave hydrothermal route and calcination at 950° C were observed [117]. In this latter study, the remanent magnetization decreased from 34 emu/g to 24 emu/g, the average grain size reduced from 660 nm to 280 nm, and the coercivity increased from 3291 Oe to 7335 Oe upon increasing x from 0.1 to 0.9. The increase in coercivity with increasing Cr content was associated with the reduction of the grain size, and the increase of the nonmagnetic α-Fe_2O_3 phase fraction, which acts as pinning centers for domain wall motion.

The substitution of Fe^{3+} by Gd^{3+} [118] or Ce^{3+} [119] was also found to improve the saturation magnetization and the coercivity of BaM ferrite. Also, the addition of 3 wt. % Bi_2O_3 to SrM hexaferrite sintered at 900° C was found to improve both the saturation magnetization and coercivity by more than double the values for the sample with no additives [120].

3.2 Fe^{3+} substitution by divalent-tetravalent metal ion combinations

For a wide range of applications, high saturation magnetization is required, but not necessarily very high coercivity. The use of M-type hexaferrites for applications such as high-density magnetic recording media requires reduction of the coercivity to $\gtrsim$ 1000 Oe without reducing the remanent magnetization, which could be satisfactory at the lower end of about 20 emu/g [121, 122]. Such properties could be achieved by special substitutions for Fe in M-type hexaferrite. Accordingly, in an effort to understand the nature of magnetic interactions, and modify the magnetic properties of hexaferrites, extensive research work had been carried out on the synthesis and characterization of M-type hexaferrites with Fe^{3+} ions substituted by special combinations of divalent–tetravalent metal ions [73, 123-142].

Specifically, Co–Zn–Nb substitution in BaM ferrite prepared by glass crystallization was found to reduce the coercivity to the range of 500 – 2000 Oe, leaving the saturation magnetization relatively high for high-density magnetic recording applications [143]. Also, it was found that Zn–Nb substitution offers an advantage over Co–Nb and Co–Ti substitutions by reducing the switching field distribution for a better high-density recording performance. In addition, Co–Sn substitution in BaM ($BaCo_xSn_xFe_{12-2x}$) prepared by a reverse micro-emulsion technique resulted in a gradual decrease of the saturation magnetization, the remanent magnetization, and coercivity with increasing x [102]. However, the saturation magnetization obtained by this method was found superior to that obtained by other chemical methods. The results of the study indicated that at x = 0.5, M_s = 70.4 emu/g, M_r = 34.0 emu/g, and H_c = 1510 Oe; these properties are suitable for high density magnetic recording.

Further, the effects on the magnetic properties of A–Sn (A = Co, Ni, Zn) substituted BaM ($BaFe_{12-2x}Sn_xA_xO_{19}$) were investigated as a function of heat treatment, substitution level, and preparation method [14]. The samples prepared by solid state reaction (SSR) were fired at 1300° C, whereas those prepared by high energy ball milling (HEM) and chemical coprecipitation (CC) methods were fired at 1000 – 1100° C and 750° C, respectively. The saturation magnetization of the Co – Sn and Ni – Sn substituted samples prepared by SSR route dropped sharply from $60 - 65$ emu/g down to $20 - 25$ emu/g as x increased from 0.5 to 1.0. The Zn – Sn substituted samples prepared by the same method, however, exhibited a small increase from ~ 53 emu/g at $x = 0.1$ to ~ 57 emu/g at $x = 1.0$, and then dropped sharply down to ~ 20 emu/g at $x = 1.5$. The samples with $x = 1.0$ prepared by HEM and CC methods exhibited higher saturation magnetizations than those prepared by SSR method, and these variations were attributed to the modification of the preferential site occupation of the substituents at this substitution level. The coercivity of the Co – Sn and Ni – Sn substituted samples prepared by SSR decreased sharply from $\sim 1100 - 1200$ Oe at $x = 0.1$ to $\sim 500 - 700$ Oe at $x = 0.5$, and then decreased at a slower rate at higher concentrations of the substituents. The Zn – Sn substituted samples, however, exhibited significantly lower coercivities in the whole concentration range. The results of the study indicated that the CC method resulted in higher coercivities, where at $x = 0.1$ the coercivities of all compounds were in the range $4140 - 4380$ Oe, characteristic of hard ferrite magnets appropriate for permanent magnet applications due to their relatively high saturation magnetization. On the other hand, at $x = 1.0$ the coercivities of the Co – Sn and Zn – Sn samples were 1040 Oe and 1180 Oe, respectively, and their saturation magnetization in the range $40 - 45$ emu/g makes them suitable for high -density magnetic recording applications. At this substitution level, however, the Ni – Sn substituted sample exhibited weaker magnetic properties with a significantly lower coercivity of 490 Oe, and a saturation magnetization below 30 emu/g.

In particular, Co–Ti substitution received a considerable interest due to the early realization of the effectiveness of this combination in reducing the coercivity down to values suitable for applications such as magnetic recording and data storage media [11, 54, 139, 144-155]. Materials with relatively high saturation magnetization and coercivities up to 1200 Oe are suitable for normal longitudinal magnetic recording, where the higher end of the coercivity is used for high density recording. Materials with higher coercivities, however, are not suitable for longitudinal recording, and could be used for high density perpendicular magnetic recording media [151, 154]. Further, potential applications of Co – Ti substituted M-type ferrites in multi-layer chip beads and other microwave applications in the hyper-frequency range were reported [156-159]. However, the work devoted to the synthesis and investigation of the magnetic properties of Co–Ti substituted SrM ferrites was extremely limited in the literature. We, therefore, devote the remaining of this article to our results concerning the synthesis and characterization of $SrFe_{12-2x}Co_xTi_xO_{19}$ ferrites prepared by ball milling and calcination.

4. Co-Ti substituted SrM ferrite

4.1 Experimental procedures

$SrFe_{12-2x}Co_xTi_xO_{19}$ (x = 0, 0.2, 0.4, 0.6, 0.8 and 1.0) precursor powders were prepared from spec pure $SrCO_3$, Fe_2O_3, TiO_2 and CoO powders (Sigma Aldrich-make). Each sample was prepared by ball-milling 8g of the starting powder mixture using a planetary ball-mill (Fritsch Pulverisette-7) with balls (10 cm in diameter) and cylindrical vial (50 cm^3) of hardened steel. The milling was carried out at 250 rpm for 16 h with ball to powder mass ratio of 8:1. The as-milled powders were annealed in air atmosphere at 1100°C for 2 h. X-ray diffraction (XRD) analysis was carried out in Philips X'Pert PRO X-ray diffractometer (PW3040/60) with CuK$_\alpha$ radiation (λ = 1.5405 Å) . XRD patterns for all fabricated samples were refined based on Rietveld analysis using FullProf suite 2000 software [160]. The particle size and morphology of the prepared samples were examined by scanning electron microscopy (SEM) (FEI Quanta 600). The magnetic measurements were carried out using vibrating sample magnetometer (VSM) (MicroMag 3900, Princeton Measurements Corporation), with a maximum applied field of 10 kOe.

4.2 XRD analysis

Fig. 1 shows XRD patterns of Co–Ti doped strontium ferrites ($SrFe_{12-2x}Co_xTi_xO_{19}$) with different doping concentrations, together with the standard pattern for $SrFe_{12}O_{19}$ (JCPDS file no: 033-1340). XRD patterns of the samples with x up to 0.8 revealed pure SrM phase ($SrFe_{12}O_{19}$) with space group P6$_3$/mmc, while a small peak at 2θ = 33.1° belonging to a minority α-Fe_2O_3 phase appeared in the sample with x = 1.0. Therefore we may conclude that Co and Ti ions with concentrations in the range reported in this work substituted Fe^{3+} ions in the $SrFe_{12}O_{19}$ lattice without affecting its hexagonal structure appreciably.

Rietveld structural refinements were carried out on XRD patterns of all fabricated samples, and the refined lattice constants and goodness of fit (indicated by χ^2 value) are shown in Table 2. A representative refined pattern is shown in Fig. 2 for the pure sample (x = 0.0). The relatively low values of χ^2 indicate good fits with reliable structural parameters. The lattice parameters for the sample with x = 0.0 agree well with those reported for strontium hexaferrite [161-163]. Also, the lattice parameters of the Co–Ti substituted samples did not change appreciably in the whole range of substitution level. These results clearly indicate that doping of strontium ferrite with Co and Ti did not induce noticeable structural changes in SrM hexaferrite.

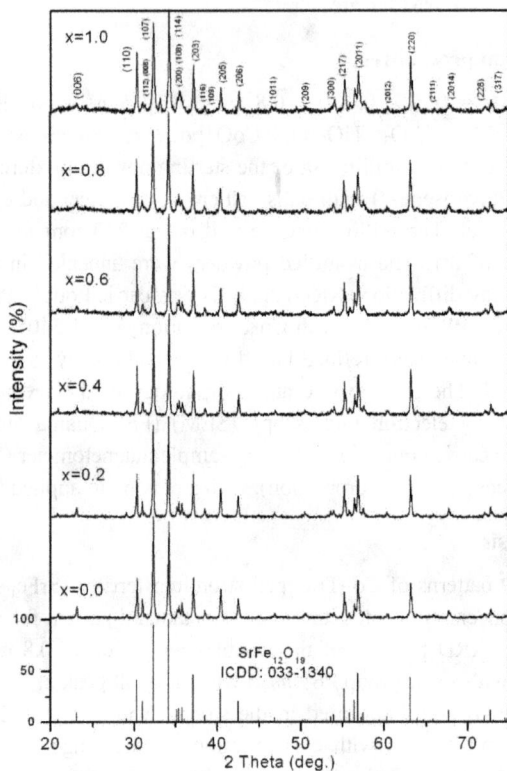

Fig. 1. XRD patterns of SrFe$_{12-2x}$Co$_x$Ti$_x$O$_{19}$ (x = 0.0 to 1.0).

The average crystallite size was determined from the (114) and (107) reflections using the method of Stokes and Wilson, where the coherence length along the direction perpendicular to the reflecting plane is given by [164].

$$D = \lambda/(\beta \cos \theta) \tag{1}$$

where D is the crystallite size, β the integral peak breadth defined as the ratio of the integrated intensity to the maximum intensity of the peak, λ the wavelength of radiation (1.54056 Å), and θ the peak position. The average crystallite size (Table 2) in the samples with x up to 0.6 remain in the narrow range of $55 - 60$ nm and only higher substitution levels show a tendency to reduce the crystallite size, which dropped down to 45 nm for the sample with $x = 1.0$.

Table 2. *Refined structural parameters and crystallite size of Co-Ti substituted SrM ferrites.*

x	χ^2	$a = b$ (Å)	c (Å)	V (Å^3)	Average crystallite size D_{XRD} (nm)
0.0	1.08	5.8814	23.0355	690.0652	57
0.2	0.90	5.8802	23.0229	689.4157	60
0.4	0.94	5.8809	23.0192	689.4473	57
0.6	1.01	5.8813	23.0170	689.4847	55
0.8	1.33	5.8826	23.0226	689.9492	49
1.0	1.63	5.8820	23.0254	690.0911	45

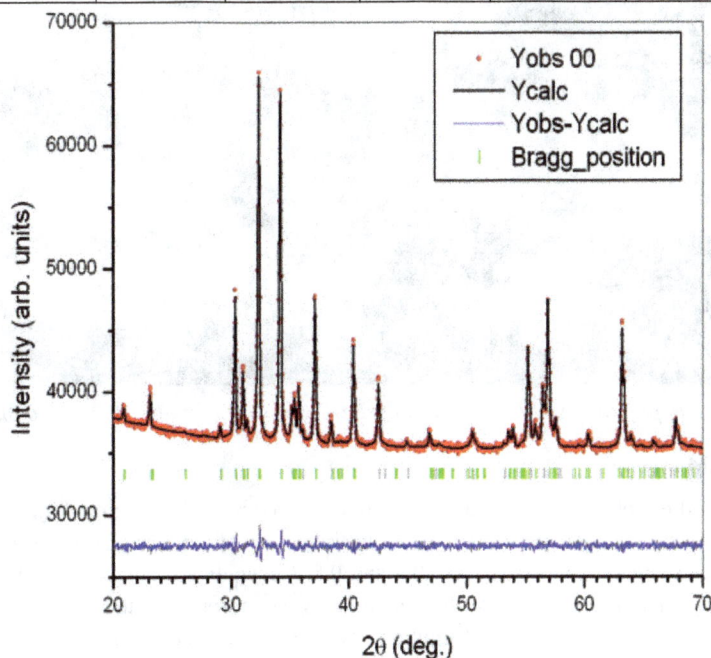

Fig. 2. *Refined XRD pattern for SrFe$_{12}$O$_{19}$.*

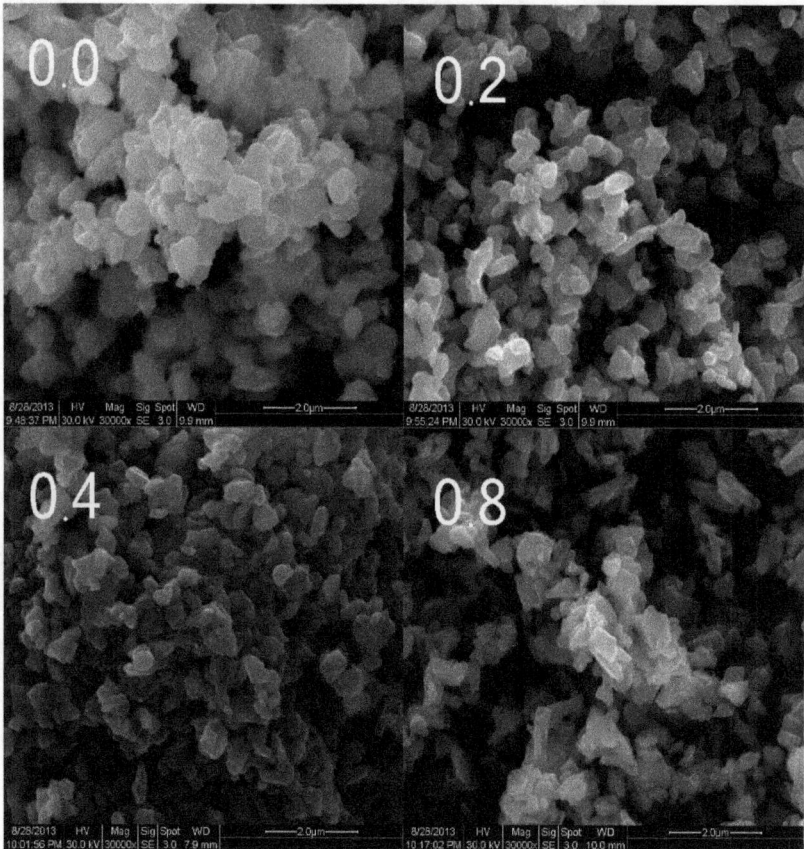

Fig. 3. Representative SEM images of $SrFe_{12-2x}Co_xTi_xO_{19}$ *(x = 0.0, 0.2, 0.4, 0.8).*

4.3 SEM imaging

The size and morphology for all samples were examined by SEM imaging. Fig. 3 shows representative SEM photographs of Co–Ti substituted strontium ferrite powders with the substituting concentration of 0.0, 0.2, 0.4 and 0.8. Generally speaking, all samples are composed of agglomerated platelet-like grains with the morphology that did not change noticeably with with Co–Ti substitution. The grain size, however, revealed small decrease with substitution, and all samples are composed of grains < 1 μm in dimension, which is within the critical single domain size [2]. There the image of the pure sample (x

= 0.0) revealed grain size generally in the range from 200 nm to 0.8 μm, while the images of the Co–Ti substituted indicated grain sized ranging from 150 nm to 0.5 μm.

4.4   Magnetism

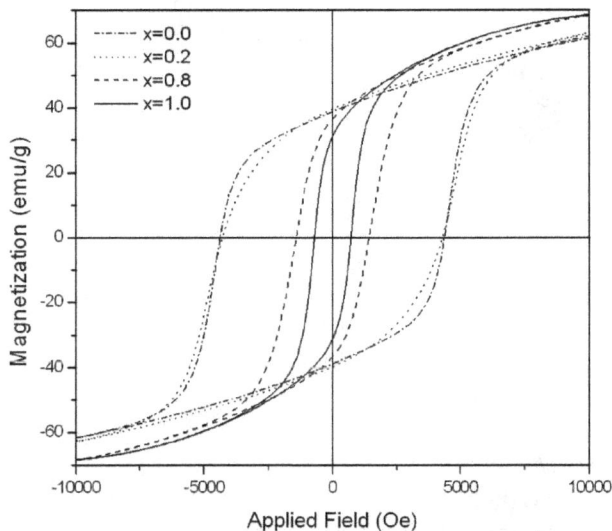

Fig. 4. Hysteresis loops for some of the SrFe$_{12-2x}$Co$_x$Ti$_x$O$_{19}$ (x = 0.0, 0.2, 0.8, 1.0) samples as a function of applied magnetic field.

Fig. 4 shows the measured hysteresis loops for representative SrFe$_{12-2x}$Co$_x$Ti$_x$O$_{19}$ samples as a function of applied magnetic field. The magnetization curve for the non-substituted sample belongs to a hard magnetic material with a high intrinsic coercivity of 4400 Oe. This value is in agreement with the previously reported values for SrM hexaferrites [115, 165]. The magnetization of all samples did not saturate at the highest applied field, and the law of approach to saturation [63, 166, 167] was employed to obtain the saturation magnetization of the samples. The effect of Co–Ti substitution on the saturation magnetization and coercivity of SrFe$_{12-2x}$Co$_x$Ti$_x$O$_{19}$ are shown in Fig. 5. With increasing x the saturation magnetization starts increasing for $x > 0.6$, recording an increase of 10% for the sample $x = 0.8$. Fig 5 also shows that the intrinsic coercivity drops dramatically from about 4400 Oe to 700 Oe for $x = 1.0$. The remanent magnetization decreased slightly from 39.5 emu/g for the sample with $x = 0.2$ down to 36.6 emu/g for the sample with $x = 0.8$, while the coercivity dropped down to 1425 Oe for the latter sample. Thus, the substitution of Fe by Co–Ti in SrM ferrite with x up to 0.8 resulted in a substantial reduction in coercivity and maintained the saturation magnetization and remanence at relatively high levels suitable for high-density magnetic recording applications.

Ferric ions in M-type hexaferrites occupy five different interstitial sites, three of which are octahedral ($12k$, $4f_2$ and $2a$), one is tetrahedral ($4f_1$) and one is trigonal bi-pyramidal ($2b$). The crystal symmetry involving Fe–O bond lengths and Fe–O–Fe bond angles between the magnetic ions in the hexaferrite lattice lead to superexchange interactions which split the magnetic structure of M-type hexaferrite into spin-up and spin-down sublattices. These sublattices and their major contributions to the magnetic properties are summarized in Table 3.

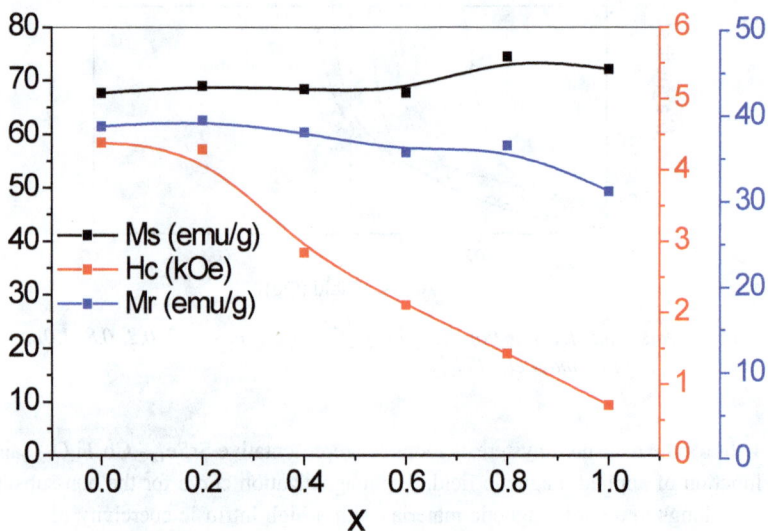

Fig. 5. Saturation magnetization, remanent magnetization, and intrinsic coercivity of $SrFe_{12-2x}Co_xTi_xO_{19}$ as a function of Co-Ti concentration (x).

Table 3. Magnetic sublattices and their contribution to the magnetic properties of M-type hexaferrites.

Sites	Coordination	Spin direction	Contribution
$4f_2$	Octahedral	Down	Magnetization; Coercivity; Anisotropy
$2b$	Bi-pyramid	Up	Magnetization; Coercivity; Anisotropy
$12k$	Octahedral	Up	Magnetization; Anisotropy
$2a$	Octahedral	Up	Magnetization
$4f_1$	Tetrahedral	Down	Magnetization

It was reported that the major positive contribution to the magnetic anisotropy in M-type hexaferrites comes from iron ions at $4f_2$ and $2b$ sites, while the contribution of $12k$ site is negative, preferring in-plane easy-axis [168]. Neutron diffraction results on Co–Ti substituted BaM indicated that 50% of the Co^{2+} ions occupy $4f_1$ site, and Ti^{4+} ions preferentially $4f_2$ site [144]. However, evidence of substitution at $12k$ and $2b$ sites was provided [168]. On the other hand, analysis of the site preference based on Mössbauer spectroscopy results indicated that the substitution occurs at $4f_2$ and $2b$ sites and that the $12k$ is not a preferred substitutional site [145]. Further, Mössbauer spectra of Co–Sn substituted BaM indicated the preference for substitution at $4f_2$, $2b$, and $12k$ sites [123]. The saturation magnetization did not change significantly with increasing x, up to 0.6, indicating that in this range of substitution, Co and Ti ions are distributed almost equally among spin-up and spin-down sublattices. For higher values of x, the slight increase in saturation magnetization could be ascribed to the involvement of the $4f_1$ spin-down sublattice as a substitutional site at high Co–Ti concentrations [145, 148, 168].

The dramatic drop of the intrinsic coercivity (Fig. 5) could not be ascribed to changes in grain size and morphology, since Co–Ti did not induce noticeable microstructural effects. This behavior is therefore associated with intrinsic parameters involving the reduction of the magnetocrystalline anisotropy. In order to investigate the effect of Co–Ti substitution on the magnetic anisotropy, the effective anisotropy field (H_a) was determined for each sample examined in this work from the switching field distribution (SFD). The switching field distribution can be obtained by differentiating the reduced IRM curve $m_r(H) = M_r(H)/M_r(\infty)$. The effective magnetic anisotropy field for each sample examined in this work is obtained from the maximum of the switching field distribution according to the formula [169]:

$$f(H)\big]_{\max} = \left[\frac{dm_r}{dH}\right]_{H=H_a/2} \tag{2}$$

Here $H_a = 2H_{max}$, where H_{max} is the value of the field at the maximum of the SFD. Fig. 6 shows representative reduced IRM curves and the corresponding switching field distributions for the samples with $x = 0.0$ and 1.0.

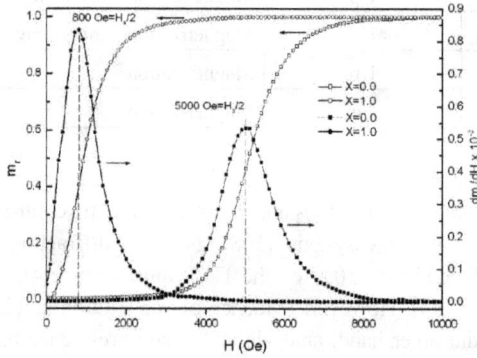

Fig. 6. Reduced IRM curves and the corresponding switching field distributions for the samples with $x = 0.0$ and $x = 1.0$.

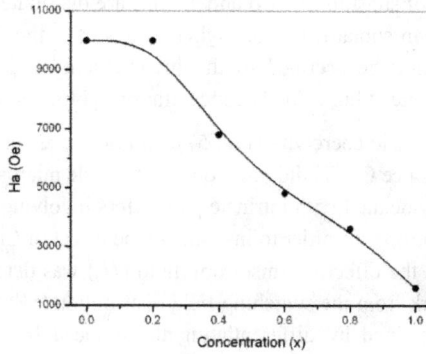

Fig. 7. Anisotropy field of $SrFe_{12-2x}Co_xTi_xO_{19}$ as a function of the Co–Ti concentration.

Fig. 7 shows the variation of magnetic anisotropy field with Co-Ti concentration for all samples examined. It is clear that H_a decreases monotonically with increasing Co-Ti concentration beyond $x = 0.2$. Considering the results of SEM imaging which revealed similarity of the particle shape and size distribution in these samples, the monotonic decline in the intrinsic coercivity for $x \geq 0.2$ is well explained by the dependence of the anisotropy field on Co–Ti concentration. This result is consistent with the reported decrease of the anisotropy constant with increasing Co–Ti, and the consequent decrease in coercivity [147].

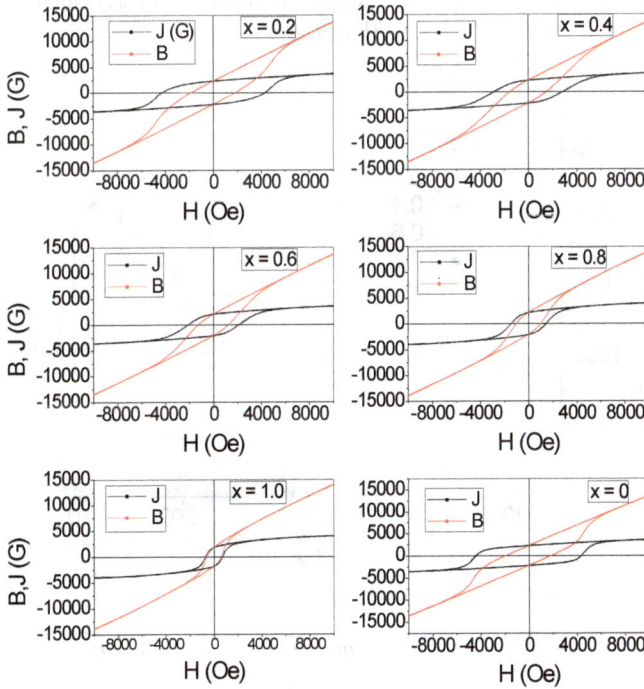

Fig. 8. B-H and J-H curves for Co–Ti substituted SrM ((x = 0.0, 0.2, 0.4, 0.6, 0.8, 1.0).

The *B-H* and *J-H* curves of the samples under investigation are shown in Fig. 8, where *J* = $4\pi M$, *M* being the magnetization in emu/cm^3. From these curves, the Retentivity (or residual induction B_r), the coercivity H_{cB}, and the intrinsic coercivity H_{cJ} can be determined. These magnetic parameters are important for evaluating the suitability of a magnetic material for a given application. Magnetic materials with high residual induction and high coercivity are sought for permanent magnet industry. For magnetic recording applications, however, the coercivity should be reduced to $\gtrsim$ 1000 Oe, while maintaining the residual induction as high as possible to provide good signal and avoid erasure of stored information by stray fields [121, 122]. The curves in Fig. 8 indicate that the sample with $x = 0$ is characterized by a linear B-H curve far into the second quadrant of the hysteresis loop, indicating that this sample is suitable as a permanent magnet with relatively high coercivity. The B-H curves deviate progressively from linearity, and the residual induction decreases with increasing Co–Ti concentration as demonstrated by the induction curves in the second quadrant (Fig. 9).

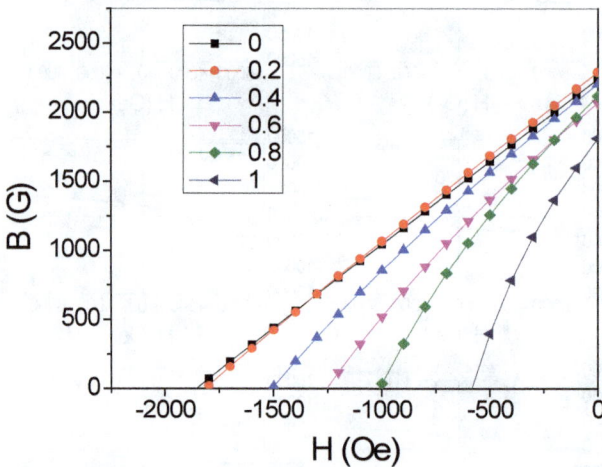

Fig. 9. B-H curves in the second quadrant for Co–Ti substituted SrM.

The maximum energy product $(BH)_{max}$ is a quality factor for a permanent magnet. A plot of BH vs. –H in the second quadrant of the hysteresis loop for the sample with $x = 0$ is shown in Fig. 10. The curve indicates that this sample is characterized by $(BH)_{max} = 1.05$ MGOe (8.35 kJ/m^3). Although this value is significantly smaller than that for RE-permanent magnets, SrM remains of practical importance for permanent magnet industry, considering the low cost of production, availability of raw materials, and corrosion resistance.

The maximum energy product, as well as the remanent induction and coercivity for all samples are evaluated from the *B-H* and *J-H* curves, and the results are summarized in Table 4. The magnetic properties of the samples indicate that these hexaferrites demonstrate a potential for permanent magnet applications in the low *x* region (up to 0.2), whereas, at higher substitution levels, the ferrites are suitable for high-density magnetic recording applications.

Table 4. Magnetic parameters of Co-Ti substituted SrM ferrites.

x	B_r (G)	H_{cB}(Oe)	H_{cJ} (Oe)	$(BH)_{max}$ (kJ/m^3)
0.0	2240	1850	4400	8.35
0.2	2288	1815	4300	8.52
0.4	2204	1500	2840	7.32
0.6	2066	1250	2200	5.84
0.8	2110	1025	1425	5.02
1.0	1800	575	620	2.62

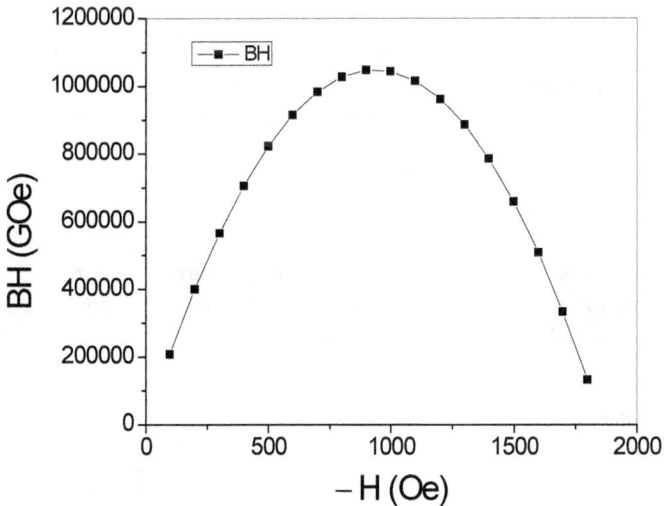

Fig. 10. BH curve in the second quadrant of the hysteresis loop the pure SrM sample.

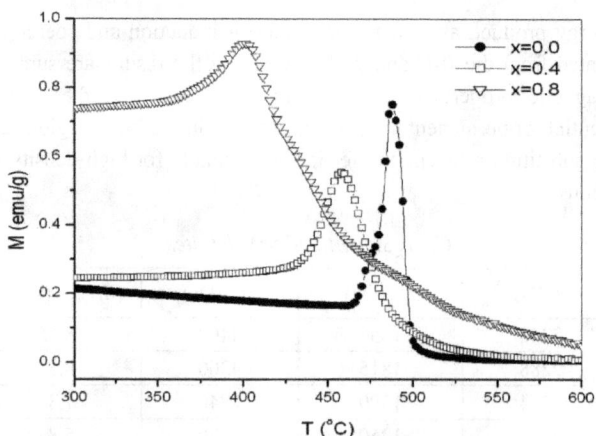

Fig. 11. Magnetization as a function of temperature for $SrFe_{12-2x}Co_xTi_xO_{19}$ with $x = 0.0$, 0.4 and 0.8.

The temperature dependence of the specific magnetization M(T) was investigated (Fig.11). The magnetization was measured as a function of temperature in a constant applied field of 100 Oe. All curves exhibited Hopkinson peak just below Curie temperature (T_c), which is characteristic of the presence of superparamagnetic grains in the samples [170]. From these curves, it is clear that the fraction of the superparamagnetic phases is reduced with Co–Ti substitution. The broadening and induced irregularity in the peak structure at high substitution levels could be due to magnetic inhomogeneity of the sample. Curie temperature as a function of Co-Ti concentration in $SrFe_{12-2x}Co_xTi_xO_{19}$ is shown in Table 5. The decrease in T_c is consistent with the conclusion that the replacement of Fe ions by Co-Ti ions results in the attenuation of the magnetic exchange coupling.

Table 5. Curie temperature for $SrFe_{12-2x}Co_xTi_xO_{19}$.

x	0.0	0.2	0.4	0.6	0.8	1.0
T_c (°C)	498	490	478	430	420	368

4.5   Inter-particle interactions

In order to investigate the role of doping with Co-Ti on the inter-particle interactions in the prepared SrM ferrites, we evaluate Δm defined by the relation [171].

$$\Delta m \left(H \right) \ = \ m_d \left(H \right) - \left[1 - 2m_r \left(H \right) \right] \tag{3}$$

Where,

$$m_r \left(H \right) = M_r \left(H \right) / M_r \left(\infty \right) \tag{4}$$

is the reduced isothermal remanent magnetization (IRM), and

$$m_d \left(H \right) = M_d \left(H \right) / M_r \left(\infty \right) \tag{5}$$

is the reduced dc demagnetization (DCD). The IRM curve was obtained by the following procedure: the sample was fist demagnetized, then a positive field is applied, and finally, the remanent magnetization is measured after removing the applied field. The procedure was repeated with increasing the positive field to reach positive saturation remanence. The DCD curve was obtained by first saturating the sample with a positive field of 10 kOe, then a negative field was applied to the sample, and the remanent magnetization was recorded after removing the negative field; this procedure was repeated with increasing the negative field until negative saturation remanence was reached.

The IRM and DCD curves for all samples examined in this work are shown in Fig. 12. $\Delta m(H)$, which gives the strength and the sign of the inter-particle interactions was evaluated for each sample using equation 3 and the data in Fig. 12. For an assembly of non-interacting particles, Δm is field-independent; any deviation from this behavior is a sign of the existence of inter-particle interactions. Positive Δm values indicate the existence of inter-particle interactions that contribute constructively to the magnetization (magnetizing-like effect), where particles tend to stack in columns. On the other hand, negative Δm values suggest that the existing interactions are demagnetizing (demagnetizing-like effect), where the particles tend to form clusters.

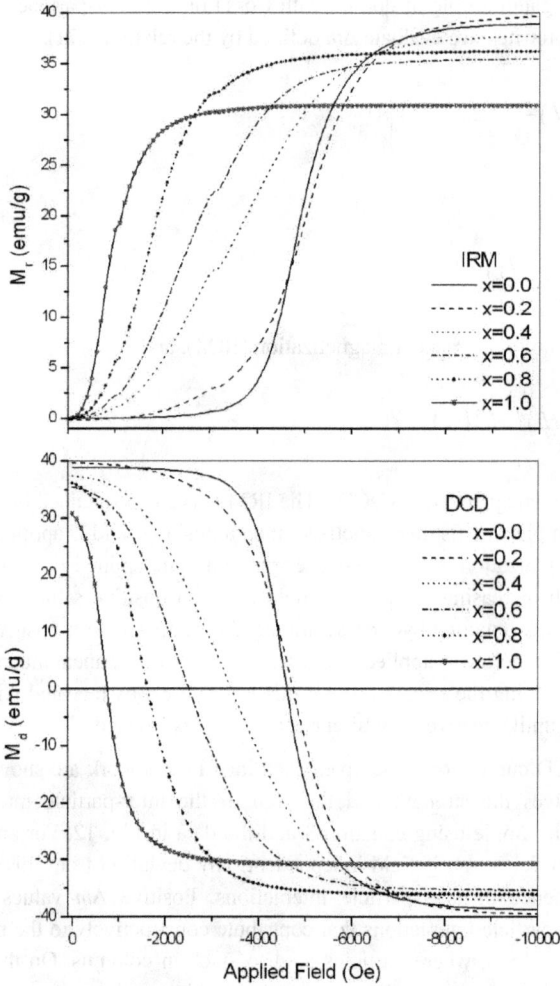

Fig. 12. IRM and DCD curves of $SrFe_{12-2x}Co_xTi_xO_{19}$.

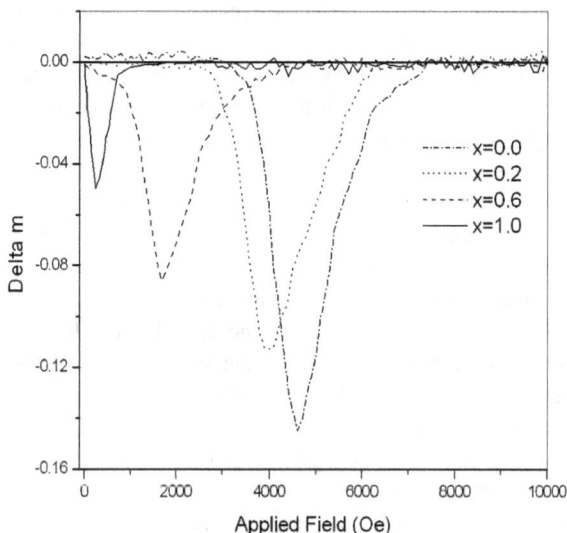

Fig. 13. Delta M curves of $SrFe_{12-2x}Co_xTi_xO_{19}$ for all concentration examined.

Fig. 13 shows the Δm curves as a function of the applied field for samples with different concentrations of Co-Ti. It is clear that Δm values for all samples are negative at all fields, with a maximum value occurring around the intrinsic coercivity. This result is an indication of a demagnetizing-like effect in all samples, in agreement with SEM observation of cluster formation. Also, the maximum negative value of Δm decreased with increasing Co-Ti concentration, indicating that Co-Ti substitution may result in the suppression of cluster formation.

5.   Conclusions

The magnetic properties of hexaferrites in current use, or those with potential for future applications, span the wide range of coercivity from few hundred Oe to several thousands of Oe, depending on the application to be used for. In this article, the enhancement of the magnetic properties of M-type hexaferrites, and modifications made by adopting different synthesis routes and experimental conditions are reviewed. Chemical methods were found to be more effective than the conventional solid state reaction in controlling the particle shapes and particle size distribution and improving the magnetic properties. Complex experimental procedures including variations of the chemical stoichiometry, heat treatment, and combinations of different synthesis techniques were adopted for further improvements of the magnetic properties of the ferrite powders. Also, the effects of the various types of substitutions on the magnetic properties of the hexaferrites were reviewed. Since the term "improvement of the properties" is device designer dependent, substitutions leading to the improvement of the magnetic properties of the ferrites for different types of device applications were addressed. Generally, the substitution of Fe ions by trivalent ions such as Al or Cr ions results in a significant increase in the coercivity suitable for permanent magnet applications. The coercivity was further enhanced by rare-earth metal substitutions. Also, the substitution of Fe^{3+} ions in M-type hexaferrites by combinations of divalent–tetravalent ions results in appreciable modifications of the magnetic properties of the hexaferrites. In particular, Co-Ti substitution was found to be effective in modifying the properties of the ferrites to be suitable for high-density magnetic recording and microwave applications. In addition, the structural and magnetic properties of a series of $SrFe_{12-2x}Co_xTi_xO_{19}$ ferrites prepared by ball milling method were carefully investigated. The effect of Co–Ti substitution on the magnetic properties was found to reduce the coercivity down to levels suitable for high density longitudinal and perpendicular magnetic recording, without reducing the saturation magnetization. Although no further improvements concerning the magnetic properties of those magnetic materials are required for such applications, control over the particle size and shape, as well as the orientation of the magnetic particles on the substrate of the recording media may require special procedures.

References

[1]   http://www.magneticsmagazine.com/main/news/permanent-magnet-market-will-reach-28-70-billion-in-2019/.

[2]   R.C. Pullar, Hexagonal ferrites: a review of the synthesis, properties and applications of hexaferrite ceramics, Progress in Materials Science, 57 (2012) 1191-1334. https://doi.org/10.1016/j.pmatsci.2012.04.001

[3]   Ü. Özgür, Y. Alivov, H. Morkoç, Microwave ferrites, part 1: fundamental properties, Journal of Materials Science: Materials in Electronics, 20 (2009) 789-834. https://doi.org/10.1007/s10854-009-9923-2

[4]   K.J. Strnat, Modern permanent magnets for applications in electro-technology, Proceedings of the IEEE, 78 (1990) 923-946. https://doi.org/10.1109/5.56908

[5] V.G. Harris, A. Geiler, Y. Chen, S.D. Yoon, M. Wu, A. Yang, Z. Chen, P. He, P.V. Parimi, X. Zuo, Recent advances in processing and applications of microwave ferrites, Journal of Magnetism and Magnetic Materials, 321 (2009) 2035-2047. https://doi.org/10.1016/j.jmmm.2009.01.004

[6] J. Smit, H.P.J. Wijn, Ferrites, Wiley, New York, 1959.

[7] S. Chikazumi, Physics of Ferromagnetism 2e, Oxford University Press 2009.

[8] S.H. Mahmood, A.N. Aloqaily, Y. Maswadeh, A. Awadallah, I. Bsoul, M. Awawdeh, H.K. Juwhari, Effects of heat treatment on the phase evolution, structural, and magnetic properties of Mo-Zn doped M-type hexaferrites, Solid State Phenomena, 232 (2015) 65-92. https://doi.org/10.4028/www.scientific.net/SSP.232.65

[9] S.H. Mahmood, M.D. Zaqsaw, O.E. Mohsen, A. Awadallah, I. Bsoul, M. Awawdeh, Q.I. Mohaidat, Modification of the magnetic properties of Co_2Y hexaferrites by divalent and trivalent metal substitutions, Solid State Phenomena, 241 (2016) 93-125. https://doi.org/10.4028/www.scientific.net/SSP.241.93

[10] G. Albanese, Recent advances in hexagonal ferrites by the use of nuclear spectroscopic methods, Le Journal de Physique Colloques, 38 (1977) 85-94. https://doi.org/10.1051/jphyscol:1977117

[11] R. Chantrell, K. O'Grady, Magnetic characterization of recording media, Journal of Physics D: Applied Physics, 25 (1992) 1. https://doi.org/10.1088/0022-3727/25/1/001

[12] H. Fu, H.R. Zhai, H.C. Zhang, B.X. Gu, J.Y. Li, Magnetic properties on Mn substituted barium ferrite, Journal of Magnetism and Magnetic Materials, 54-57 (1986) 905-906. https://doi.org/10.1016/0304-8853(86)90307-0

[13] I.Y. Gershov, Barium ferrite permanent magnets, Soviet Powder Metallurgy and Metal Ceramics, 1 (1964) 386-393. https://doi.org/10.1007/BF00774124

[14] D. Lisjak, M. Drofenik, Synthesis and characterization of A–Sn-substituted (A= Zn, Ni, Co) BaM–hexaferrite powders and ceramics, Journal of the European Ceramic Society, 24 (2004) 1841-1845. https://doi.org/10.1016/S0955-2219(03)00445-X

[15] S. Mahmood, A. Aloqaily, Y. Maswadeh, A. Awadallah, I. Bsoul, H. Juwhari, Structural and Magnetic Properties of Mo-Zn Substituted ($BaFe_{12-4x}Mo_xZn_{3x}O_{19}$) M-Type Hexaferrites, Material Science Research India, 11 (2014) 09-20.

[16] G. Turilli, F. Licci, S. Rinaldi, A. Deriu, Mn^{2+}, Ti^{4+} substituted barium ferrite, Journal of Magnetism and Magnetic Materials, 59 (1986) 127-131. https://doi.org/10.1016/0304-8853(86)90019-3

[17] A. Awadallah, S.H. Mahmood, Y. Maswadeh, I. Bsoul, M. Awawdeh, Q.I. Mohaidat, H. Juwhari, Structural, magnetic, and Mossbauer spectroscopy of Cu substituted M-type hexaferrites, Materials Research Bulletin, 74 (2016) 192-201. https://doi.org/10.1016/j.materresbull.2015.10.034

[18] O.T. Ozkan, H. Erkalfa, The effect of B_2O_3 addition on the direct sintering of barium hexaferrite, Journal of the European Ceramic Society, 14 (1994) 351-358. https://doi.org/10.1016/0955-2219(94)90072-8

[19] P. Hernandez-Gomez, J.M. Munoz, C. Torres, C. de Francisco, O. Alejos, Influence of stoichiometry on the magnetic disaccommodation in barium M-type hexaferrites, Journal of Physics D: Applied Physics, 36 (2003) 1062-1070. https://doi.org/10.1088/0022-3727/36/9/303

[20] Y. Maswadeh, S.H. Mahmood, A. Awadallah, A.N. Aloqaily, Synthesis and structural characterization of nonstoichiometric barium hexaferrite materials with Fe: Ba ratio of 11.5–16.16, IOP Conference Series: Materials Science and Engineering, IOP Publishing, 2015, pp. 012019.

[21] Y.-M. Kang, Y.-H. Kwon, M.-H. Kim, D.-Y. Lee, Enhancement of magnetic properties in Mn–Zn substituted M-type Sr-hexaferrites, Journal of Magnetism and Magnetic Materials, 382 (2015) 10-14. https://doi.org/10.1016/j.jmmm.2015.01.048

[22] Y. Maswadeh, Structural analysis of hexaferrite materials, Physics, The University of Jordan, 2014.

[23] P. Garcia-Casillas, A. Beesley, D. Bueno, J. Matutes-Aquino, C. Martinez, Remanence properties of barium hexaferrite, Journal of alloys and compounds, 369 (2004) 185-189. https://doi.org/10.1016/j.jallcom.2003.09.100

[24] D. Lisjak, M. Drofenik, The mechanism of the low-temperature formation of barium hexaferrite, Journal of the European Ceramic Society, 27 (2007) 4515-4520. https://doi.org/10.1016/j.jeurceramsoc.2007.02.202

[25] J.-P. Wang, L. Ying, M.-L. Zhang, Y.-j. QIAO, X. Tian, Comparison of the Sol-gel Method with the Coprecipitation Technique for Preparation of Hexagonal Barium Ferrite, Chemical Research in Chinese Universities, 24 (2008) 525-528. https://doi.org/10.1016/S1005-9040(08)60110-5

[26] V. Harikrishnan, P. Saravanan, R.E. Vizhi, D.R. Babu, V. Vinod, P. Kejzlar, M. Černík, Effect of annealing temperature on the structural and magnetic properties of CTAB-capped $SrFe_{12}O_{19}$ platelets, Journal of Magnetism and Magnetic Materials, 401 (2016) 775-783. https://doi.org/10.1016/j.jmmm.2015.10.122

[27] H.B. von Basel, K.A. Hempel, Static magnetic properties of pressure-sintered barium ferrite, Journal of Magnetism and Magnetic Materials, 38 (1983) 316-318. https://doi.org/10.1016/0304-8853(83)90373-6

[28] S.E. Jacobo, C. Domingo-Pascual, R. Rodrigez-Clemente, M.A. Blesa, Synthesis of ultrafine particles of barium ferrite by chemical coprecipitation, Journal of Materials Science, 33 (1997) 1025-1028. https://doi.org/10.1023/A:1018582423406

[29] M. Rashad, I. Ibrahim, A novel approach for synthesis of M-type hexaferrites nanopowders via the co-precipitation method, Journal of Materials Science: Materials in Electronics, 22 (2011) 1796-1803. https://doi.org/10.1007/s10854-011-0365-2

[30] A. Davoodi, B. Hashemi, Investigation of the effective parameters on the synthesis of strontium hexaferrite nanoparticles by chemical coprecipitation method, Journal of Alloys and Compounds, 512 (2012) 179-184. https://doi.org/10.1016/j.jallcom.2011.09.059

[31] S.R. Janasi, D. Rodrigues, F.J. Landgraf, M. Emura, Magnetic properties of coprecipitated barium ferrite powders as a function of synthesis conditions, Magnetics, IEEE Transactions on, 36 (2000) 3327-3329. https://doi.org/10.1109/INTMAG.2000.872434

[32] J. Matutes-Aquino, S. Dıaz-Castanón, M. Mirabal-Garcıa, S. Palomares-Sánchez, Synthesis by coprecipitation and study of barium hexaferrite powders, Scripta Materialia, 42 (2000) 295-299. https://doi.org/10.1016/S1359-6462(99)00350-4

[33] P. Shepherd, K.K. Mallick, R.J. Green, Magnetic and structural properties of M-type barium hexaferrite prepared by co-precipitation, Journal of Magnetism and Magnetic Materials, 311 (2007) 683-692. https://doi.org/10.1016/j.jmmm.2006.08.046

[34] Z. Mosleh, P. Kameli, A. Poorbaferani, M. Ranjbar, H. Salamati, Structural, magnetic and microwave absorption properties of Ce-doped barium hexaferrite, Journal of Magnetism and Magnetic Materials, 397 (2016) 101-107. https://doi.org/10.1016/j.jmmm.2015.08.078

[35] M. Jamalian, An investigation of structural, magnetic and microwave properties of strontium hexaferrite nanoparticles prepared by a sol–gel process with doping SN and Tb, Journal of Magnetism and Magnetic Materials, 378 (2015) 217-220. https://doi.org/10.1016/j.jmmm.2014.11.047

[36] W. Zhong, W. Ding, N. Zhang, J. Hong, Q. Yan, Y. Du, Key step in synthesis of ultrafine $BaFe_{12}O_{19}$ by sol-gel technique, Journal of Magnetism and Magnetic Materials, 168 (1997) 196-202. https://doi.org/10.1016/S0304-8853(96)00664-6

[37] R.C. Alange, P.P. Khirade, S.D. Birajdar, A.V. Humbe, K.M. Jadhav, Structural, magnetic and dielectric properties of Al-Cr co-substituted M-type barium hexaferrite nanoparticles, Journal of Molecular Structure, 1106 (2016) 460-467. https://doi.org/10.1016/j.molstruc.2015.11.004

[38] Y. Hong, C. Ho, H.Y. Hsu, C. Liu, Synthesis of nanocrystalline $Ba(MnTi)_xFe_{12-2x}O_{19}$ powders by the sol–gel combustion method in citrate acid–metal nitrates system ($x = 0$, 0.5, 1.0, 1.5, 2.0), Journal of Magnetism and Magnetic Materials, 279 (2004) 401-410. https://doi.org/10.1016/j.jmmm.2004.02.008

[39] S.H. Mahmood, F.S. Jaradat, A.F. Lehlooh, A. Hammoudeh, Structural properties and hyperfine interactions in Co-Zn Y-type hexaferrites prepared by sol-gel method, Ceramics International, 40 (2014) 5231-5236. https://doi.org/10.1016/j.ceramint.2013.10.092

[40] W. Abbas, I. Ahmad, M. Kanwal, G. Murtaza, I. Ali, M.A. Khan, M.N. Akhtar, M. Ahmad, Structural and magnetic behavior of Pr-substituted M-type hexagonal ferrites synthesized by sol–gel autocombustion for a variety of applications, Journal of Magnetism and Magnetic Materials, 374 (2015) 187-191. https://doi.org/10.1016/j.jmmm.2014.08.029

[41] Simon Thompson, Neil J. Shirtcliffe, Eoin S. O'Keefe, Steve Appleton, C.C. Perry, Synthesis of $SrCo_xTi^xFe_{(12-2x)}O_{19}$ through sol-gel auto-ignition and its characterisation, Journal of Magnetism and Magnetic Materials, 297 (2005) 100-1007. https://doi.org/10.1016/j.jmmm.2004.10.102

[42] Y. Meng, M. He, Q. Zeng, D. Jiao, S. Shukla, R. Ramanujan, Z. Liu, Synthesis of barium ferrite ultrafine powders by a sol–gel combustion method using glycine gels, Journal of Alloys and Compounds, 583 (2014) 220-225. https://doi.org/10.1016/j.jallcom.2013.08.156

[43]   D. Bahadur, S. Rajakumar, A. Kumar, Influence of fuel ratios on auto combustion synthesis of barium ferrite nano particles, Journal of Chemical Sciences, 118 (2006) 15-21. https://doi.org/10.1007/BF02708761

[44]   V. Sankaranarayanan, Q. Pankhurst, D. Dickson, C. Johnson, Ultrafine particles of barium ferrite from a citrate precursor, Journal of Magnetism and Magnetic Materials, 120 (1993) 73-75. https://doi.org/10.1016/0304-8853(93)91290-N

[45]   V. Sankaranarayanan, D. Khan, Mechanism of the formation of nanoscale M-type barium hexaferrite in the citrate precursor method, Journal of Magnetism and Magnetic Materials, 153 (1996) 337-346. https://doi.org/10.1016/0304-8853(95)00537-4

[46]   V. Sankaranarayanan, Q. Pankhurst, D. Dickson, C. Johnson, An investigation of particle size effects in ultrafine barium ferrite, Journal of Magnetism and Magnetic Materials, 125 (1993) 199-208. https://doi.org/10.1016/0304-8853(93)90838-S

[47]   X. Liu, J. Wang, L.-M. Gan, S.-C. Ng, Improving the magnetic properties of hydrothermally synthesized barium ferrite, Journal of Magnetism and Magnetic Materials, 195 (1999) 452-459. https://doi.org/10.1016/S0304-8853(99)00123-7

[48]   A. Ataie, I. Harris, C. Ponton, Magnetic properties of hydrothermally synthesized strontium hexaferrite as a function of synthesis conditions, Journal of Materials Science, 30 (1995) 1429-1433. https://doi.org/10.1007/BF00375243

[49]   D. Primc, D. Makovec, D. Lisjak, M. Drofenik, Hydrothermal synthesis of ultrafine barium hexaferrite nanoparticles and the preparation of their stable suspensions, Nanotechnology, 20 (2009) 315605. https://doi.org/10.1088/0957-4484/20/31/315605

[50]   M. Drofenik, I. Ban, D. Makovec, A. Žnidaršič, Z. Jagličić, D. Hanžel, D. Lisjak, The hydrothermal synthesis of super-paramagnetic barium hexaferrite particles, Materials Chemistry and Physics, 127 (2011) 415-419. https://doi.org/10.1016/j.matchemphys.2011.02.037

[51]   R.H. Arendt, The molten salt synthesis of single domain $BaFe_{12}O_{19}$ and $SrFe_{12}O_{19}$ crystals, Journal of Solid State Chemistry, 8 (1973) 339-347. https://doi.org/10.1016/S0022-4596(73)80031-3

[52]   T.-S. Chin, S. Hsu, M. Deng, Barium ferrite particulates prepared by a salt-melt method, Journal of Magnetism and Magnetic Materials, 120 (1993) 64-68. https://doi.org/10.1016/0304-8853(93)91288-I

[53] Y. Liu, M.G. Drew, Y. Liu, J. Wang, M. Zhang, Preparation, characterization and magnetic properties of the doped barium hexaferrites $BaFe_{12-2x}Co_{x/2}Zn_{x/2}Sn_xO_{19}$, x = 0.0–2.0, Journal of Magnetism and Magnetic Materials, 322 (2010) 814-818. https://doi.org/10.1016/j.jmmm.2009.11.009

[54] O. Kubo, T. Ido, H. Yokoyama, Properties of Ba ferrite particles for perpendicular magnetic recording media, Magnetics, IEEE Transactions on, 18 (1982) 1122-1124. https://doi.org/10.1109/TMAG.1982.1062007

[55] O. Kubo, T. Ido, T. Nomura, K. Inomata, Method for manufacturing magnetic powder for high density magnetic recording, Google Patents, 1982.

[56] B. Shirk, W. Buessem, Magnetic properties of barium ferrite formed by crystallization of a glass, Journal of the American Ceramic Society, 53 (1970) 192-196. https://doi.org/10.1111/j.1151-2916.1970.tb12069.x

[57] D. Jung, S. Hong, J. Cho, Y. Kang, Nano-sized barium titanate powders with tetragonal crystal structure prepared by flame spray pyrolysis, Journal of the European Ceramic Society, 28 (2008) 109-115. https://doi.org/10.1016/j.jeurceramsoc.2007.05.018

[58] J.S. Cho, D.S. Jung, S.K. Hong, Y.C. Kang, Characteristics of nano-sized pb-based glass powders by high temperature spray pyrolysis method, Journal of the Ceramic Society of Japan, 116 (2008) 600-604. https://doi.org/10.2109/jcersj2.116.600

[59] D.-H. Kim, Y.-K. Lee, K.-M. Kim, K.-N. Kim, S.-Y. Choi, I.-B. Shim, Synthesis of Ba-ferrite microspheres doped with Sr for thermoseeds in hyperthermia, Journal of Materials Science, 39 (2004) 6847-6850. https://doi.org/10.1023/B:JMSC.0000045617.92955.12

[60] M.H. Kim, D.S. Jung, Y.C. Kang, J.H. Choi, Nanosized barium ferrite powders prepared by spray pyrolysis from citric acid solution, Ceramics International, 35 (2009) 1933-1937. https://doi.org/10.1016/j.ceramint.2008.10.016

[61] U. Topal, Improvement of the remanence properties and the weakening of interparticle interactions in $BaFe_{12}O_{19}$ particles by B_2O_3 addition, Physica B: Condensed Matter, 407 (2012) 2058-2062. https://doi.org/10.1016/j.physb.2012.02.004

[62] U. Topal, Towards Further Improvements of the Magnetization Parameters of B_2O_3-Doped $BaFe_{12}O_{19}$ Particles: Etching with Hydrochloric Acid, Journal of

Superconductivity and Novel Magnetism, 25 (2012) 1485-1488. https://doi.org/10.1007/s10948-012-1486-4

[63]   A. Awadallah, S.H. Mahmood, Y. Maswadeh, I. Bsoul, A. Aloqaily, Structural and magnetic properties of Vanadium Doped M-Type Barium Hexaferrite ($BaFe_{12-x}V_xO_{19}$), IOP Conference Series: Materials Science and Engineering, IOP Publishing, 2015, pp. 012006. https://doi.org/10.1088/1757-899X/92/1/012006

[64]   Q. Mohsen, Barium hexaferrite synthesis by oxalate precursor route, Journal of Alloys and Compounds, 500 (2010) 125-128. https://doi.org/10.1016/j.jallcom.2010.03.230

[65]   T. Gonzalez-Carreno, M. Morales, C. Serna, Barium ferrite nanoparticles prepared directly by aerosol pyrolysis, Materials Letters, 43 (2000) 97-101. https://doi.org/10.1016/S0167-577X(99)00238-4

[66]   U. Topal, H. Ozkan, H. Sozeri, Synthesis and characterization of nanocrystalline $BaFe_{12}O_{19}$ obtained at 850 C by using ammonium nitrate melt, Journal of Magnetism and Magnetic Materials, 284 (2004) 416-422. https://doi.org/10.1016/j.jmmm.2004.07.009

[67]   U. Topal, H. Ozkan, L. Dorosinskii, Finding optimal Fe/Ba ratio to obtain single phase $BaFe_{12}O_{19}$ prepared by ammonium nitrate melt technique, Journal of Alloys and Compounds, 428 (2007) 17-21. https://doi.org/10.1016/j.jallcom.2006.03.047

[68]   S. El-Sayed, T. Meaz, M. Amer, H. El Shersaby, Magnetic behavior and dielectric properties of aluminum substituted M-type barium hexaferrite, Physica B: Condensed Matter, 426 (2013) 137-143. https://doi.org/10.1016/j.physb.2013.06.026

[69]   V.V. Soman, V. Nanoti, D. Kulkarni, Dielectric and magnetic properties of Mg–Ti substituted barium hexaferrite, Ceramics International, 39 (2013) 5713-5723. https://doi.org/10.1016/j.ceramint.2012.12.089

[70]   H. Sözeri, Effect of pelletization on magnetic properties of $BaFe_{12}O_{19}$, Journal of Alloys and Compounds, 486 (2009) 809-814. https://doi.org/10.1016/j.jallcom.2009.07.072

[71]   I. Bsoul, S. Mahmood, Structural and magnetic properties of $BaFe_{12-x}Al_xO_{19}$ prepared by milling and calcination, Jordan Journal of Physics, 2 (2009) 171-179.

[72]   I. Bsoul, S. Mahmood, Magnetic and structural properties of $BaFe_{12-x}Ga_xO_{19}$ nanoparticles, Journal of Alloys and Compounds, 489 (2010) 110-114. https://doi.org/10.1016/j.jallcom.2009.09.024

[73]   I. Bsoul, S. Mahmood, A.-F. Lehlooh, Structural and magnetic properties of $BaFe_{12-2x}Ti_xRu_xO_{19}$, Journal of Alloys and Compounds, 498 (2010) 157-161. https://doi.org/10.1016/j.jallcom.2010.03.142

[74]   H.-F. Yu, $BaFe_{12}O_{19}$ powder with high magnetization prepared by acetone-aided coprecipitation, Journal of Magnetism and Magnetic Materials, 341 (2013) 79-85. https://doi.org/10.1016/j.jmmm.2013.04.030

[75]   Y. Liu, M.G. Drew, Y. Liu, Preparation and magnetic properties of barium ferrites substituted with manganese, cobalt, and tin, Journal of Magnetism and Magnetic Materials, 323 (2011) 945-953. https://doi.org/10.1016/j.jmmm.2010.11.075

[76]   E. Pashkova, E. Solovyova, I. Kotenko, T. Kolodiazhnyi, A. Belous, Effect of preparation conditions on fractal structure and phase transformations in the synthesis of nanoscale M-type barium hexaferrite, Journal of Magnetism and Magnetic Materials, 323 (2011) 2497-2503. https://doi.org/10.1016/j.jmmm.2011.05.026

[77]   G. Litsardakis, I. Manolakis, C. Serletis, K. Efthimiadis, High coercivity Gd-substituted Ba hexaferrites, prepared by chemical coprecipitation, Journal of Applied Physics, 103 (2008) 07E501.

[78]   W. Roos, H. Haak, C. Voigt, K. Hempel, Microwave absorption and static magnetic properties of coprecipitated barium ferrite, Le Journal de Physique Colloques, 38 (1977) C1-35-C31-37.

[79]   S. Kanagesan, M. Hashim, S. Jesurani, T. Kalaivani, I. Ismail, Influence of Zn–Nb on the Magnetic Properties of Barium Hexaferrite, Journal of Superconductivity and Novel Magnetism, 27 (2014) 811-815. https://doi.org/10.1007/s10948-013-2357-3

[80]   T. Kaur, A. Srivastava, Effect of pH on Magnetic Properties of Doped Barium Hexaferrite, International Journal of Research in Mechanical Engineering & Technology, 3 (2013) 171-173.

[81]   F. Khademi, A. Poorbafrani, P. Kameli, H. Salamati, Structural, magnetic and microwave properties of Eu-doped barium hexaferrite powders, Journal of Superconductivity and Novel Magnetism, 25 (2012) 525-531. https://doi.org/10.1007/s10948-011-1323-1

[82]   Y. Li, Q. Wang, H. Yang, Synthesis, characterization and magnetic properties on nanocrystalline $BaFe_{12}O_{19}$ ferrite, Current Applied Physics, 9 (2009) 1375-1380. https://doi.org/10.1016/j.cap.2009.03.002

[83] C. Sürig, D. Bonnenberg, K. Hempel, P. Karduck, H. Klaar, C. Sauer, Effects of Variations in Stoichiometry on M-Type Hexaferrites, Le Journal de Physique IV, 7 (1997) C1-315-C311-316.

[84] V.C. Chavan, S.E. Shirsath, M.L. Mane, R.H. Kadam, S.S. More, Transformation of hexagonal to mixed spinel crystal structure and magnetic properties of Co^{2+} substituted $BaFe_{12}O_{19}$, Journal of Magnetism and Magnetic Materials, 398 (2016) 32-37. https://doi.org/10.1016/j.jmmm.2015.09.002

[85] R.K. Mudsainiyan, A.K. Jassal, M. Gupta, S.K. Chawla, Study on structural and magnetic properties of nanosized M-type Ba-hexaferrites synthesized by urea assisted citrate precursor route, Journal of Alloys and Compounds, 645 (2015) 421-428. https://doi.org/10.1016/j.jallcom.2015.04.218

[86] H. Sözeri, Z. Durmuş, A. Baykal, E. Uysal, Preparation of high quality, single domain $BaFe_{12}O_{19}$ particles by the citrate sol–gel combustion route with an initial Fe/Ba molar ratio of 4, Materials Science and Engineering: B, 177 (2012) 949-955. https://doi.org/10.1016/j.mseb.2012.04.023

[87] V.N. Dhage, M. Mane, M. Babrekar, C. Kale, K. Jadhav, Influence of chromium substitution on structural and magnetic properties of $BaFe_{12}O_{19}$ powder prepared by sol–gel auto combustion method, Journal of Alloys and Compounds, 509 (2011) 4394-4398. https://doi.org/10.1016/j.jallcom.2011.01.040

[88] M. Han, Y. Ou, W. Chen, L. Deng, Magnetic properties of Ba-M-type hexagonal ferrites prepared by the sol–gel method with and without polyethylene glycol added, Journal of Alloys and Compounds, 474 (2009) 185-189. https://doi.org/10.1016/j.jallcom.2008.06.047

[89] T. Yamauchi, Y. Tsukahara, T. Sakata, H. Mori, T. Chikata, S. Katoh, Y. Wada, Barium ferrite powders prepared by microwave-induced hydrothermal reaction and magnetic property, Journal of Magnetism and Magnetic Materials, 321 (2009) 8-11. https://doi.org/10.1016/j.jmmm.2008.07.005

[90] S. Dursun, R. Topkaya, N. Akdoğan, S. Alkoy, Comparison of the structural and magnetic properties of submicron barium hexaferrite powders prepared by molten salt and solid state calcination routes, Ceramics International, 38 (2012) 3801-3806. https://doi.org/10.1016/j.ceramint.2012.01.028

[91] G. Albanese, A. Deriu, Magnetic properties of Al, Ga, Sc, In substituted barium ferrites: a comparative analysis, Ceramics International, 5 (1979) 3-10. https://doi.org/10.1016/0390-5519(79)90002-4

[92] M.H. Shams, A.S. Rozatian, M.H. Yousefi, J. Valíček, V. Šepelák, Effect of Mg^{2+} and Ti^{4+} dopants on the structural, magnetic and high-frequency ferromagnetic properties of barium hexaferrite, Journal of Magnetism and Magnetic Materials, 399 (2016) 10-18. https://doi.org/10.1016/j.jmmm.2015.08.099

[93] R. Pullar, A. Bhattacharya, The magnetic properties of aligned M hexa-ferrite fibres, Journal of Magnetism and Magnetic Materials, 300 (2006) 490-499. https://doi.org/10.1016/j.jmmm.2005.06.001

[94] A. Alsmadi, I. Bsoul, S. Mahmood, G. Alnawashi, F. Al-Dweri, Y. Maswadeh, U. Welp, Magnetic study of M-type Ru-Ti doped strontium hexaferrite nanocrystalline particles, Journal of Alloys and Compounds, 648 (2015) 419-427. https://doi.org/10.1016/j.jallcom.2015.06.274

[95] R. Palomino, A.B. Miró, F. Tenorio, F.S. De Jesús, C.C. Escobedo, S. Ammar, Sonochemical assisted synthesis of $SrFe_{12}O_{19}$ nanoparticles, Ultrasonics Sonochemistry, 29 (2016) 470-475. https://doi.org/10.1016/j.ultsonch.2015.10.023

[96] A. Bolarín-Miró, F. Sánchez-De Jesús, C.A. Cortes-Escobedo, S. Diaz-De La Torre, R. Valenzuela, Synthesis of M-type $SrFe_{12}O_{19}$ by mechanosynthesis assisted by spark plasma sintering, Journal of Alloys and Compounds, 643 (2015) S226-S230. https://doi.org/10.1016/j.jallcom.2014.11.124

[97] S. Singhal, T. Namgyal, J. Singh, K. Chandra, S. Bansal, A comparative study on the magnetic properties of $MFe_{12}O_{19}$ and $MAlFe_{11}O_{19}$ (M= Sr, Ba and Pb) hexaferrites with different morphologies, Ceramics International, 37 (2011) 1833-1837. https://doi.org/10.1016/j.ceramint.2011.02.001

[98] A. Guerrero-Serrano, T. Pérez-Juache, M. Mirabal-García, J. Matutes-Aquino, S. Palomares-Sánchez, Effect of barium on the properties of lead hexaferrite, Journal of Superconductivity and Novel Magnetism, 24 (2011) 2307-2312. https://doi.org/10.1007/s10948-011-1181-x

[99] M.N. Ashiq, R.B. Qureshi, M.A. Malana, M.F. Ehsan, Synthesis, structural, magnetic and dielectric properties of zirconium copper doped M-type calcium strontium hexaferrites, Journal of Alloys and Compounds, 617 (2014) 437-443. https://doi.org/10.1016/j.jallcom.2014.08.015

[100] P. Popa, E. Rezlescu, C. Doroftei, N. Rezlescu, Influence of calcium on properties of strontium and barium ferrites for magnetic media prepared by combustion, J. Optoelectronics Advance Materials, 7 (2005) 1553-1556.

[101] Ashima, S. Sanghi, A. Agarwal, Reetu, Rietveld refinement, electrical properties and magnetic characteristics of Ca–Sr substituted barium hexaferrites, Journal of Alloys and Compounds, 513 (2012) 436-444. https://doi.org/10.1016/j.jallcom.2011.10.071

[102] X. Gao, Y. Du, X. Liu, P. Xu, X. Han, Synthesis and characterization of Co–Sn substituted barium ferrite particles by a reverse microemulsion technique, Materials Research Bulletin, 46 (2011) 643-648. https://doi.org/10.1016/j.materresbull.2011.02.002

[103] C.-J. Li, B. Wang, J.-N. Wang, Magnetic and microwave absorbing properties of electrospun $Ba_{(1-x)}La_xFe_{12}O_{19}$ nanofibers, Journal of Magnetism and Magnetic Materials, 324 (2012) 1305-1311. https://doi.org/10.1016/j.jmmm.2011.11.016

[104] Y.-M. Kang, High saturation magnetization in La–Ce–Zn–doped M-type Sr-hexaferrites, Ceramics International, 41 (2015) 4354-4359. https://doi.org/10.1016/j.ceramint.2014.11.125

[105] L. Peng, L. Li, R. Wang, Y. Hu, X. Tu, X. Zhong, Microwave sintered $Sr_{1-x}La_xFe_{12-x}Co_xO_{19}$ ($x = $ 0–0.5) ferrites for use in low temperature co-fired ceramics technology, Journal of Alloys and Compounds, 656 (2016) 290-294. https://doi.org/10.1016/j.jallcom.2015.08.263

[106] S. Ounnunkad, Improving magnetic properties of barium hexaferrites by La or Pr substitution, Solid State Communications, 138 (2006) 472-475. https://doi.org/10.1016/j.ssc.2006.03.020

[107] M. Awawdeh, I. Bsoul, S.H. Mahmood, Magnetic properties and Mössbauer spectroscopy on Ga, Al, and Cr substituted hexaferrites, Journal of Alloys and Compounds, 585 (2014) 465-473. https://doi.org/10.1016/j.jallcom.2013.09.174

[108] S. Wang, J. Ding, Y. Shi, Y. Chen, High coercivity in mechanically alloyed $BaFe_{10}Al_2O_{19}$, Journal of Magnetism and Magnetic Materials, 219 (2000) 206-212. https://doi.org/10.1016/S0304-8853(00)00450-9

[109] I. Ali, M. Islam, M. Awan, M. Ahmad, Effects of Ga–Cr substitution on structural and magnetic properties of hexaferrite ($BaFe_{12}O_{19}$) synthesized by sol–gel auto-combustion route, Journal of Alloys and Compounds, 547 (2013) 118-125. https://doi.org/10.1016/j.jallcom.2012.08.122

[110] Joonghoe Dho, E.K. Lee, N.H.H. J.Y. Park, Effects of the grain boundary on the coercivity of barium ferrite $BaFe_{12}O_{19}$, Journal of Magnetism and Magnetic Materials, 285 (2005) 164-168. https://doi.org/10.1016/j.jmmm.2004.07.033

[111] J. Dahal, L. Wang, S. Mishra, V. Nguyen, J. Liu, Synthesis and magnetic properties of $SrFe_{12-x-y}Al_xCo_yO_{19}$ nanocomposites prepared via autocombustion technique, Journal of Alloys and Compounds, 595 (2014) 213-220. https://doi.org/10.1016/j.jallcom.2013.12.186

[112] B. Rai, S. Mishra, V. Nguyen, J. Liu, Synthesis and characterization of high coercivity rare-earth ion doped $Sr_{0.9}RE_{0.1}Fe_{10}Al_2O_{19}$ (RE: Y, La, Ce, Pr, Nd, Sm, and Gd), Journal of Alloys and Compounds, 550 (2013) 198-203. https://doi.org/10.1016/j.jallcom.2012.09.021

[113] P. Kazin, L. Trusov, D. Zaitsev, Y.D. Tretyakov, M. Jansen, Formation of submicron-sized $SrFe_{12-x}Al_xO_{19}$ with very high coercivity, Journal of Magnetism and Magnetic Materials, 320 (2008) 1068-1072. https://doi.org/10.1016/j.jmmm.2007.10.020

[114] D. Chen, Y. Liu, Y. Li, K. Yang, H. Zhang, Microstructure and magnetic properties of Al-doped barium ferrite with sodium citrate as chelate agent, Journal of Magnetism and Magnetic Materials, 337 (2013) 65-69. https://doi.org/10.1016/j.jmmm.2013.02.036

[115] A.A. Nourbakhsh, M. Noorbakhsh, M. Nourbakhsh, M. Shaygan, K.J. Mackenzie, The effect of nano sized $SrFe_{12}O_{19}$ additions on the magnetic properties of chromium-doped strontium-hexaferrite ceramics, Journal of Materials Science: Materials in Electronics, 22 (2011) 1297-1302. https://doi.org/10.1007/s10854-011-0303-3

[116] S. Ounnunkad, P. Winotai, Properties of Cr-substituted M-type barium ferrites prepared by nitrate–citrate gel-autocombustion process, Journal of Magnetism and Magnetic Materials, 301 (2006) 292-300. https://doi.org/10.1016/j.jmmm.2005.07.003

[117] S. Katlakunta, S.S. Meena, S. Srinath, M. Bououdina, R. Sandhya, K. Praveena, Improved magnetic properties of Cr^{3+} doped $SrFe_{12}O_{19}$ synthesized via microwave hydrothermal route, Materials Research Bulletin, 63 (2015) 58-66. https://doi.org/10.1016/j.materresbull.2014.11.043

[118] V.P. Singh, G. Kumar, R. Kotnala, J. Shah, S. Sharma, K. Daya, K.M. Batoo, M. Singh, Remarkable magnetization with ultra-low loss $BaGd_xFe_{12-x}O_{19}$ nanohexaferrites for applications up to C-band, Journal of Magnetism and Magnetic Materials, 378 (2015) 478-484. https://doi.org/10.1016/j.jmmm.2014.11.071

[119] R. Pawar, S. Desai, Q. Tamboli, S.E. Shirsath, S. Patange, Ce^{3+} incorporated structural and magnetic properties of M type barium hexaferrites, Journal of Magnetism and Magnetic Materials, 378 (2015) 59-63. https://doi.org/10.1016/j.jmmm.2014.10.166

[120] P. Long, H. Yue-Bin, G. Cheng, L. Le-Zhong, W. Rui, H. Yun, T. Xiao-Qiang, Preparation and magnetic properties of $SrFe_{12}O_{19}$ ferrites suitable for use in self-biased LTCC circulators, Chinese Physics Letters, 32 (2015) 017502. https://doi.org/10.1088/0256-307X/32/1/017502

[121] G. Bate, Magnetic recording materials since 1975, Journal of Magnetism and Magnetic materials, 100 (1991) 413-424. https://doi.org/10.1016/0304-8853(91)90831-T

[122] G. Bate, Recording materials, in: P. E, Wohlfarth (Ed.) Ferromagnetic materials, North-Holland Publishing Company, New York, 1980, pp. 381-508.

[123] D. Han, Z. Yang, H. Zeng, X. Zhou, A. Morrish, Cation site preference and magnetic properties of Co-Sn-substituted Ba ferrite particles, Journal of Magnetism and Magnetic Materials, 137 (1994) 191-196. https://doi.org/10.1016/0304-8853(94)90205-4

[124] A. Gonzalez-Angeles, G. Mendoza-Suarez, A. Gruskova, I. Toth, V. Jančárik, M. Papanova, J. Escalante-Garcí, Magnetic studies of NiSn-substituted barium hexaferrites processed by attrition milling, Journal of Magnetism and Magnetic Materials, 270 (2004) 77-83. https://doi.org/10.1016/j.jmmm.2003.08.001

[125] A. González-Angeles, G. Mendoza-Suárez, A. Grusková, J. Sláma, J. Lipka, M. Papánová, Magnetic structure of $Sn^{2+}Ru^{4+}$-substituted barium hexaferrites prepared by mechanical alloying, Materials Letters, 59 (2005) 1815-1819. https://doi.org/10.1016/j.matlet.2005.01.072

[126] A. González-Angeles, G. Mendoza-Suarez, A. Grusková, M. Papanova, J. Slama, Magnetic studies of Zn–Ti-substituted barium hexaferrites prepared by mechanical milling, Materials Letters, 59 (2005) 26-31. https://doi.org/10.1016/j.matlet.2004.09.012

[127] A. González-Angeles, G. Mendoza-Suarez, A. Grusková, J. Lipka, M. Papanova, J. Slama, Effect of (Ni, Zn) Ru mixtures on magnetic properties of barium hexaferrites yielded by high-energy milling, Journal of Magnetism and Magnetic Materials, 285 (2005) 450-455. https://doi.org/10.1016/j.jmmm.2004.08.015

[128] I. Bsoul, S.H. Mahmood, A.F. Lehlooh, A. Al-Jamel, Structural and magnetic properties of $SrFe_{12-2x}Ti_xRu_xO_{19}$, Journal of Alloys and Compounds, 551 (2013) 490-495. https://doi.org/10.1016/j.jallcom.2012.11.062

[129] G.H. Dushaq, S.H. Mahmood, I. Bsoul, H.K. Juwhari, B. Lahlouh, M.A. AlDamen, Effects of molybdenum concentration and valence state on the structural and magnetic properties of $BaFe_{11.6}Mo_xZn_{0.4-x}O_{19}$ hexaferrites, Acta Metallurgica Sinica (English Letters), 26 (2013) 509-516. https://doi.org/10.1007/s40195-013-0075-2

[130] S.H. Mahmood, G.H. Dushaq, I. Bsoul, M. Awawdeh, H.K. Juwhari, B.I. Lahlouh, M.A. AlDamen, Magnetic Properties and Hyperfine Interactions in M-Type $BaFe_{12-2x}Mo_xZn_xO_{19}$ Hexaferrites, Journal of Applied Mathematics and Physics, 2 (2014) 77-87. https://doi.org/10.4236/jamp.2014.25011

[131] H. Vincent, E. Brando, B. Sugg, Cationic Distribution in Relation to the Magnetic Properties of New M-Hexaferrites with Planar Magnetic Anisotropy $BaFe_{12-2x}Ir_xMe_xO_{19}$ (Me= Co, Zn, $x \approx 0.85$ and $x \approx 0.50$), Journal of Solid State Chemistry, 120 (1995) 17-22. https://doi.org/10.1006/jssc.1995.1369

[132] B. Sugg, H. Vincent, Magnetic properties of new M-type hexaferrites $BaFe_{12-2x}Ir_xCo_xO_{19}$, Journal of Magnetism and Magnetic Materials, 139 (1995) 364-370. https://doi.org/10.1016/0304-8853(95)90016-0

[133] M.V. Rane, D. Bahadur, S. Mandal, M. Patni, Characterization of $BaFe_{12-2x}Co_xZr_xO_{19}$ ($0 \leq x \leq 0.5$) synthesised by citrate gel precursor route, Journal of Magnetism and Magnetic Materials, 153 (1996) L1-L4. https://doi.org/10.1016/0304-8853(95)00305-3

[134] S. Sugimoto, K. Okayama, S.-i. Kondo, H. Ota, M. Kimura, Y. Yoshida, H. Nakamura, D. Book, T. Kagotani, M. Homma, Barium M-type ferrite as an electromagnetic microwave absorber in the GHz range, Materials Transactions, JIM, 39 (1998) 1080-1083.

[135] H. Fang, Z. Yang, C. Ong, Y. Li, C. Wang, Preparation and magnetic properties of (Zn–Sn) substituted barium hexaferrite nanoparticles for magnetic recording, Journal of Magnetism and Magnetic Materials, 187 (1998) 129-135. https://doi.org/10.1016/S0304-8853(98)00139-5

[136] P. Wartewig, M. Krause, P. Esquinazi, S. Rösler, R. Sonntag, Magnetic properties of Zn-and Ti-substituted barium hexaferrite, Journal of Magnetism and Magnetic Materials, 192 (1999) 83-99. https://doi.org/10.1016/S0304-8853(98)00382-5

[137] F. Wei, H. Fang, C. Ong, C. Wang, Z. Yang, Magnetic properties of $BaFe_{12-2x}Zn_xZr_xO_{19}$ particles, Journal of Applied Physics, 87 (2000) 8636-8639. https://doi.org/10.1063/1.373589

[138] G. Mendoza-Suarez, L. Rivas-Vazquez, J. Corral-Huacuz, A. Fuentes, J. Escalante-Garcí, Magnetic properties and microstructure of $BaFe_{11.6-2x}Ti_xM_xO_{19}$ (M= Co, Zn, Sn) compounds, Physica B: Condensed Matter, 339 (2003) 110-118. https://doi.org/10.1016/j.physb.2003.08.120

[139] A.M. Alsmadi, I. Bsoul, S.H. Mahmood, G. Alnawashi, K. Prokeš, K. Siemensmeyer, B. Klemke, H. Nakotte, Magnetic study of M-type doped barium hexaferrite nanocrystalline particles, Journal of Applied Physics, 114 (2013) 243910. https://doi.org/10.1063/1.4858383

[140] Z. Yang, C. Wang, X. Li, H. Zeng, (Zn, Ni, Ti) substituted barium ferrite particles with improved temperature coefficient of coercivity, Materials Science and Engineering: B, 90 (2002) 142-145. https://doi.org/10.1016/S0921-5107(01)00925-4

[141] S. Pignard, H. Vincent, E. Flavin, F. Boust, Magnetic and electromagnetic properties of RuZn and RuCo substituted $BaFe_{12}O_{19}$, Journal of Magnetism and Magnetic Materials, 260 (2003) 437-446. https://doi.org/10.1016/S0304-8853(02)01387-2

[142] S. Nilpairach, W. Udomkichdaecha, I. Tang, Coercivity of the co-precipitated prepared hexaferrites, $BaFe_{12-2x}Co_xSn_xO_{19}$, Journal of the Korean Physical Society, 48 (2006) 939-945.

[143] O. Kubo, E. Ogawa, Barium ferrite particles for high density magnetic recording, Journal of Magnetism and Magnetic Materials, 134 (1994) 376-381. https://doi.org/10.1016/0304-8853(94)00147-2

[144] X. Batlle, X. Obradors, J. Rodriguez-Carvajal, M. Pernet, M. Cabanas, M. Vallet, Cation distribution and intrinsic magnetic properties of Co-Ti-doped M-type barium ferrite, Journal of Applied Physics, 70 (1991) 1614-1623. https://doi.org/10.1063/1.349526

[145] X. Zhou, A. Morrish, Z. Li, Y. Hong, Site preference for Co^{2+} and Ti^{4+} in Co-Ti substituted barium ferrite, IEEE Transactions on Magnetics, 27 (1991) 4654-4656. https://doi.org/10.1109/20.278906

[146] A. Gruskova, J. Slama, M. Michalikova, J. Lipka, I. Toth, P. Kaboš, Preparation of substituted barium ferrite powders, Journal of Magnetism and Magnetic Materials, 101 (1991) 227-229. https://doi.org/10.1016/0304-8853(91)90738-V

[147] A. Morrish, X. Zhou, Z. Yang, H.-X. Zeng, Substituted barium ferrites; sources of anisotropy, Hyperfine Interactions, 90 (1994) 365-369. https://doi.org/10.1007/BF02069140

[148] Z. Šimša, S. Lego, R. Gerber, E. Pollert, Cation distribution in Co-Ti-substituted barium hexaferrites: a consistent model, Journal of Magnetism and Magnetic Materials, 140 (1995) 2103-2104. https://doi.org/10.1016/0304-8853(94)01393-4

[149] G. Bottoni, Magnetization stability and interactions in particulate recording media, Materials Chemistry and Physics, 42 (1995) 45-50. https://doi.org/10.1016/0254-0584(95)01551-5

[150] K. Kakizaki, N. Hiratsuka, T. Namikawa, Fine structure of acicular $BaCo_xTi_xFe_{12-2x}O_{19}$ particles and their magnetic properties, Journal of Magnetism and Magnetic Materials, 176 (1997) 36-40. https://doi.org/10.1016/S0304-8853(97)00634-3

[151] Y. Li, R. Liu, Z. Zhang, C. Xiong, Synthesis and characterization of nanocrystalline $BaFe_{9.6}Co_{0.8}Ti_{0.8}M_{0.8}O_{19}$ particles, Materials Chemistry and Physics, 64 (2000) 256-259. https://doi.org/10.1016/S0254-0584(99)00218-7

[152] G. Mendoza-Suarez, J. Corral-Huacuz, M. Contreras-García, H. Juarez-Medina, Magnetic properties of $BaFe_{11.6-2x}Co_xTi_xO_{19}$ particles produced by sol–gel and spray-drying, Journal of Magnetism and Magnetic Materials, 234 (2001) 73-79. https://doi.org/10.1016/S0304-8853(01)00286-4

[153] M. Kuznetsov, Q. Pankhurst, I. Parkin, Novel SHS routes to CoTi-doped M-type ferrites, Journal of Materials Science: Materials in Electronics, 12 (2001) 533-536. https://doi.org/10.1023/A:1012405610723

[154] A. Gruskova, J. Slama, R. Dosoudil, D. Kevicka, V. Jančárik, I. Toth, Influence of Co–Ti substitution on coercivity in Ba ferrites, Journal of Magnetism and Magnetic Materials, 242 (2002) 423-425. https://doi.org/10.1016/S0304-8853(01)01139-8

[155] S. Y. An, I.-B. Shim, C.S. Kim, Mössbauer and magnetic properties of Co–Ti substituted barium hexaferrite nanoparticles, Journal of Applied Physics, 91 (2002) 8465-8467. https://doi.org/10.1063/1.1452203

[156] C. Wang, L. Li, J. Zhou, X. Qi, Z. Yue, High-frequency magnetic properties of Co-Ti substituted barium ferrites prepared by modified chemical coprecipitation

method, Journal of Materials Science: Materials in Electronics, 13 (2002) 713-716. https://doi.org/10.1023/A:1021560920450

[157] C. Wang, X. Qi, L. Li, J. Zhou, X. Wang, Z. Yue, High-frequency magnetic properties of low-temperature sintered Co-Ti substituted barium ferrites, Materials Science and Engineering: B, 99 (2003) 270-273. https://doi.org/10.1016/S0921-5107(02)00521-4

[158] Z. Haijun, L. Zhichao, M. Chenliang, Y. Xi, Z. Liangying, W. Mingzhong, Preparation and microwave properties of Co-and Ti-doped barium ferrite by citrate sol–gel process, Materials Chemistry and Physics, 80 (2003) 129-134. https://doi.org/10.1016/S0254-0584(02)00457-1

[159] R. Lima, M.S. Pinho, M.L. Gregori, R.R. Nunes, T. Ogasawara, Effect of double substituted *m*-barium hexaferrites on microwave absorption properties, Materials Science-Poland, 22 (2004) 245-252.

[160] J. Rodriguez-Carvajal, FULLPROF 98. Program for Rietveld Pattern Matching Analysis of Powder Patterns, unpublished results, Grenoble, 1998.(b) Rodriguez-Carvajal, J, Physica B, 55 (1993) 192.

[161] X. Obradors, X. Solans, A. Collomb, D. Samaras, J. Rodriguez, M. Pernet, M. Font-Altaba, Crystal structure of strontium hexaferrite $SrFe_{12}O_{19}$, Journal of Solid State Chemistry, 72 (1988) 218-224. https://doi.org/10.1016/0022-4596(88)90025-4

[162] O. Kalogirou, G. Haack, B. Röhl, W. Gunßer, Mössbauer study of a modified M-type Ba (Sr)-ferrite prepared by ion exchange, Solid State Ionics, 63 (1993) 528-533. https://doi.org/10.1016/0167-2738(93)90156-W

[163] G. B. Teh, Y.C. Wong, R.D. Tilley, Effect of annealing temperature on the structural, photoluminescence and magnetic properties of sol–gel derived Magnetoplumbite-type (M-type) hexagonal strontium ferrite, Journal of Magnetism and Magnetic Materials, 323 (2011) 2318-2322. https://doi.org/10.1016/j.jmmm.2011.04.014

[164] B. E. Warren, X-ray Diffraction, Addison-Wesley, Reading, Massachsetts, 1969.

[165] S. Masoudpanah, S.S. Ebrahimi, Fe/Sr ratio and calcination temperature effects on processing of nanostructured strontium hexaferrite thin films by a sol–gel method, Research on Chemical Intermediates, 37 (2011) 259-266. https://doi.org/10.1007/s11164-011-0286-y

[166] B. D. Cullity, C.D. Graham, Introduction to magnetic materials, John Wiley & Sons, 2011.

[167] S. H. Mahmood, A. Awadallah, Y. Maswadeh, I. Bsoul, Structural and magnetic properties of Cu-V substituted M-type barium hexaferrites, IOP Conference Series: Materials Science and Engineering, IOP Publishing, 2015, pp. 012008. https://doi.org/10.1088/1757-899X/92/1/012008

[168] Q. Pankhurst, Anisotropy field measurement in barium ferrite powders by applied field Mossbauer spectroscopy, Journal of Physics: Condensed Matter, 3 (1991) 1323. https://doi.org/10.1088/0953-8984/3/10/010

[169] H. Pfeiffer, W. Schüppel, Investigation of magnetic properties of barium ferrite powders by remanence curves, Physica Status Solidi (a), 119 (1990) 259-269. https://doi.org/10.1002/pssa.2211190131

[170] S. H. Mahmood, I. Bsoul, Hopkinson peak and superparamagnetic effects in $BaFe_{12-x}Ga_xO_{19}$ nanoparticles, EPJ Web of Conferences, 29 (2012) 00039.

[171] P. Kelly, K. O'Grady, P. Mayo, R. Chantrell, Switching mechanisms in cobalt-phosphorus thin films, IEEE Transactions on Magnetics, 25 (1989) 3881-3883. https://doi.org/10.1109/20.42466

Chapter 3

Influence of (Glycine/Nitrate) Ratio on Structural and the Magnetic Properties of $Gd_3Fe_5O_{12}$

S.I. El-Dek [1,a], S.F. Mansour[2,b], N. Okasha[3,c] and M.A. Ahmed[4,d]

[1]Materials Science and Nanotechnology Deptartment, Faculty of Post graduate studies for Advanced Sciences, (PSAS),Beni-Suef University, Beni-Suef, Egypt

[2]Physics Department, Faculty of Science, Zagazig University, Zagazig, Egypt

[3]Physics Department, Faculty of Girls, Ain Shams University, Cairo, Egypt

[4]Materials Science Laboratory (1), Physics Department, Faculty of Science, Cairo University, Giza, Egypt

[a]didi5550000@gmail.com, [a]samaa@psas.bsu.edu.eg, [b]salws_fahim@yahoo.com, [c]nagwa.okasha@women.asu.edu.eg, [d]moala47@hotmail.com,

Abstract

Gadolinium iron garnet (GdIG) $Gd_3Fe_5O_{12}$ was prepared using the autocombustion method with glycine as fuel. The GdIG samples revealed single phase garnet with cubic symmetry. The effect of (glycine/ nitrate) ratio on the structural and magnetic properties of the investigated garnet is reported. The results of the study indicated that the lattice parameter decreased with a remarkable improvement of the densification with increasing (glycine/ nitrate) ratio.

Keywords

GdIG Nanoparticles, Glycine/Nitrate Ratio, XRD, TEM, Magnetization

Contents

1.   Introduction

Garnets have a cubic crystal structure of oxygen polyhedra, with the trivalent rare earth and Fe^{3+} metal cations occupying tetrahedral (D), octahedral (A), or dodecahedral - (C) sites. Specifically, the exchange interactions between Fe^{3+} ions at the tetrahedral and octahedral sites promote antiparallel alignment of the moments of the Fe^{3+} ions, with net magnetic moment antiparallel to that of the rare earth ions on the C sites. The garnet structure is one of the most complicated crystals and it is difficult to draw a two-dimensional representation that shows clearly all the ions (160) in the unit cell [1-3].

Rare earth iron garnets (RIG) are promising candidates for the use in high-performance microwaves and electrochemical devices owing to their high resistivity, high Curie temperature, high chemical stability, and they possess unique magnetic, optical, thermo-physical, and mechanical properties [4-10]. The general chemical formula for the RIG is $\{R^{3+}_3\}_C [Fe^{3+}_2]_A (Fe^{3+}_3)_D O^{2-}_{12\,H}$. Its unit cell is composed of eight formula units forming a complex cubic lattice consisting of 160 ions on specific lattice positions, i.e., 96 O^{2-} ions on H sites, and three groups of trivalent metal ions distributed on 24 C sites $\{R^{3+}_3\}$, 16 octahedral A sites $[Fe^{3+}_2]$ and 24 tetrahedral D sites (Fe^{3+}_3), respectively.

The desired properties for specific applications of gadolinium iron garnet (GdIG) [11-14] have been provided by controlling the preparation conditions or by the variation of appropriate substitutions in the garnet. In recent years, many researchers have focused on the study of the dependence of the physical and chemical properties on the particle size of the materials [11, 12].

However, there is no report in the literature, to the best of our knowledge, on the synthesis of single phase nanostructured GdIG with various crystal sizes. In the past few decades, attempts have been made to synthesize nanostructured GdIG using various techniques and when the size of the particles is reduced to nanoscale, it was found that the garnet decomposes into rare earth ortho-ferrite, and other rare earth and iron oxide phases [15, 16].

The combustion technique received much attention because of its capability to produce ultrafine powder and pure phase at low temperatures revealing high density at comparatively lower temperatures [17-19]. This technique involves an exothermic decomposition of the metal nitrates forming a gel which helps in preventing phase separation whereas the heat of combustion reaction is utilized for the phase formation.

Amino Acid glycine (NH_2CH_2COOH) is capable of complexing a number of metal ions of different ionic sizes including rare earth and transition metals. The glycine/nitrate ratio used in the combustion reaction plays a predominant role in the nature of the reaction and the powder properties [20].

In the present work, we report the successful synthesis of nanocrystalline single phase of (GdIG) using the flash autocombustion method at low annealing temperature. Another goal of our work is to investigate the influence of the variation of glycine /nitrate ratio (G/N) on the magnetic properties of GdIG.

2. Experimental procedure

Stoichiometric mixtures of gadolinium nitrate; ($Gd(NO_3)_3$. $6H_2O$), iron nitrate; (Fe $(NO_3)_3$. $9H_2O$), and glycine as a fuel were used as raw materials to prepare GdIG sample via modified flash autocombustion technique (MFAT) [21]. These raw materials are weighed according to the $Gd_3Fe_5O_{12}$ stoichiometry using different ratios of glycine to nitrate (G/N) namely; 0.5:1, 0.75:1, 1:1, and 1.66:1 respectively. The mixture was mixed well on a magnetic stirrer until a homogeneous solution is achieved. After that, the mixture was dried at 353 K on a hot plate for few mins, and finally annealed at 1150°C with heating rate 10°C/min for 2h in air using a Lenton furnace (16/5 UAF England), then cooled to room temperature at the same rate as that of the heating. The final products were crushed again in an agate mortar to fine powder.

Thermogravimetric analysis (TGA) using analyzer (Shimadzu TGA- 50H) was carried out for the as synthesized powder to monitor the decomposition process of the samples at a heating rate of 10°C/min in N_2 atmosphere. The crystal structure, was investigated using an X- ray diffractometer (Proker D_8 - USA) using CuKα radiation with wavelength λ = 1.5418 Å. The formation of GdIG phase was verified by comparing the XRD pattern of the nanoparticles with the ICDD card (74-1361). The crystallite size (L) was calculated from the full width at half maximum of a Bragg reflection using Debye-Scherrer's formula [22];

$$L = 0.9 \, \lambda \, / \beta \cos \theta \tag{1}$$

where, λ is the wavelength of the X- ray, β is the full width at half-maximum (FWHM) and θ is the Bragg's angle. For carrying the transmission electron microscope (TEM), the powder particles were dispersed in ethylene glycol and then the ultra sonicator is used for homogenization. TEM was performed using JEOL, JEM – 2100. The dc molar magnetic susceptibility (χ_M) of the investigated samples was measured using Faraday's method for the powdered samples at magnetic field intensity 1990 Oe and different applied temperatures ranging between 300- 700 K, where the sample was inserted at the point of maximum gradient. The temperature of the sample was detected using copper constantan thermocouple connected to Digi-sense thermometer (USA) with junction in contact with the sample. The accuracy of measuring temperature was better than ±1°C.

3. Results and discussion

3.1 Structure charcterization

To investigate the formation of the garnet phase, thermal analysis for the as synthesized samples was carried out from room temperature up to 1200° C in N_2-atmosphere using heating rate of 10° C min^{-1}. Fig. 1 shows the typical TGA/Dr (TGA) curve for the sample with (G/N) ratio = 0.5. Different weight loss steps are observed in the thermo gram. From TGA and on continuous heating, the decomposition of nitrates in the metal complex takes place with the evolution of gases like NO_2, NO, or N_2H_5. Weight losses were observed at 150, 325 and 780°C. The weight loss which begins at 70°C and was completed at 150°C corresponds to the loss of water molecules of the started nitrates.

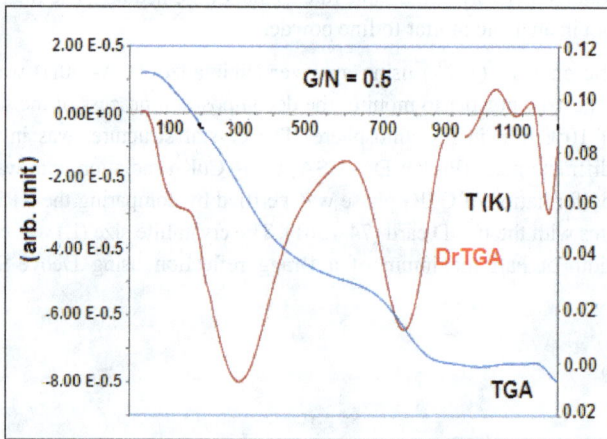

Fig. 1. Thermo gravimetric analysis of the GdIG sample at G/N=0.5.

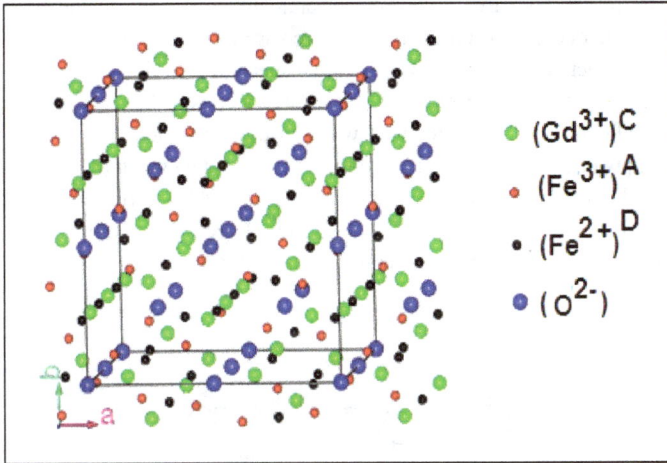

Fig. 2. Schematic crystal structure of the GdIG garnet showing the rare earth ionic position at dodecahedral (C) sites, and Fe^{3+} positions at Octahedral (A) and tetrahedral (D) sites.

The weight loss at 325°C corresponds to the loss of organic residues such as COOH. The weight loss at 780°C corresponds to the crystallization of GdIG due to oxidation of metal glycine complexes [23, 24]. This peak shifts to lower values with increasing (G/N) ratio due to increasing the fuel ratio which in turns vary the ignition temperature. The last peak at 1179°C corresponds to crystallization of the mixed oxides of gadolinium and iron to obtain crystalline gadolinium- iron garnet (GdIG) single phase. This peak (phase formation) shifts also depending on the (G/N) ratio. The adopted annealing temperature of 1150°C in the present work was based on the results of TGA analysis as it represents the starting temperature of complete phase formation.

The crystal structure of the GdIG crystal (G/N = 0.5) is produced by E.P.Cryst [25] from the obtained diffraction data (Fig. 1) and presented in Fig. 2.

The X-ray diffraction pattern of the investigated samples is shown in Fig. 3. The XRD data of the samples were compared and indexed with ICDD card (74-1361). The samples with (G/N) ratios = 0.5, 0.75 and 1 revealed a cubic garnet phase structure with the appearance of a very small intensity hematite phase. This small amount (less than 5%) could be neglected and the small intensity corresponds to iron oxide (Fe_2O_3) in the sample of (G/N) ratio = 1.66. The lattice parameter was computed according to the cubic symmetry. The initial decrease in the lattice parameter (a) is due to the following: i) Shrinkage of the lattice as a result of the cation redistribution and mechanically induced stresses. ii) With increasing (G/N) ratio, possible mechanically induced changes in the bond lengths and angles are expected. The calculated values of lattice parameter (a) are listed in Table (1) and found to agree well with those reported for GdIG samples ICDD card (74-1361) [14]. The crystallite size (L) was determined by using the Debye-Scherrer formula from the full width at half maximum of the (420) diffraction peak and it decreased with increasing (G/N) ratio, Table 1. This decrease is possibly related to the completion of the reaction at lower temperature with increasing fuel.

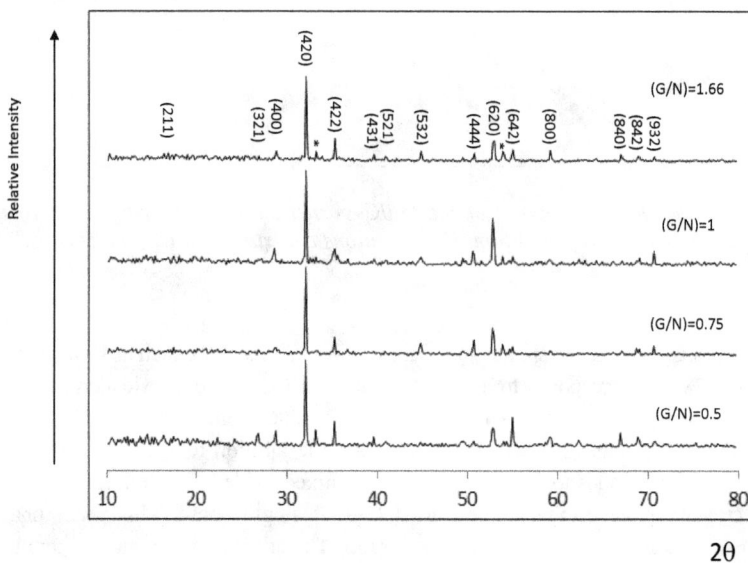

Fig. 3. XRD pattern of the GdIG samples prepared at different (G/N) ratios

This assumption is supported by the shift of the peak at 780°C with increasing (G/N) ratio to lower temperatures in the TGA thermograms. Fig. 1. The X-ray density was calculated from $D_x=ZM/NV$ where Z is the number of molecules per unit cell, M is the molecular weight, N is Avogadro s number and V is the unit cell volume and reported in Table 1. The density increased with (G/N) ratio due to the decrease in the unit cell volume.

Table 1. The values of the phase formation temperature, intensity of the secondary phase, crystallite size L (nm), the lattice parameter a (Å), theoretical density D_x (gcm^{-3}) as a function of (G/N) ratio for the GdIG samples.

(G/N) ratio	Phase formation temp. ($^{\circ}$C)	Fe$_2$O$_3$ Peak intensity %	L (nm)	a (Å)	D_x (gcm^{-3})
0.5	1179	7	228	12.4689	6.46
0.75	1170	4	156	12.4671	6.46
1	1165	3.5	143	12.4627	6.47
1.66	1159	6	134	12.4516	6.49

Fig. 4 a-d shows TEM micrographs of the investigated samples with different (G/N) ratios. The micrographs show agglomerates of particles. However, the particle size and nature of agglomeration among them was found to be sensitive to the (G/N) ratio used for the synthesis process. The formation of such agglomerates is due to the high flame temperature generated during the preparation. The particles are likely seen to be elongated in shape which certainly would affect the magnetocrystalline anisotropy of the nanogarnet. Moreover, with increasing (G/N) ratio the agglomeration of the particles as well as the size decreased and the morphology changed.

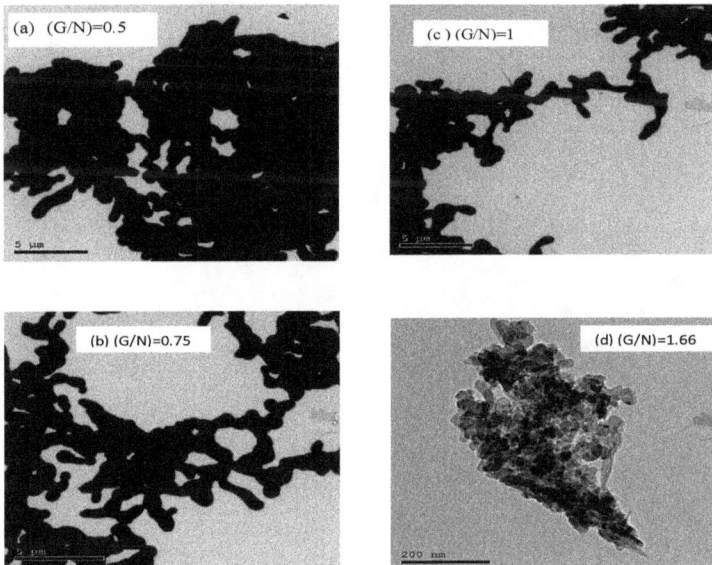

Fig. 4 a-d. Transmission electron micrographs of GdIG samples prepared at different (G/N) ratios.

3.2   Magnetic Properties

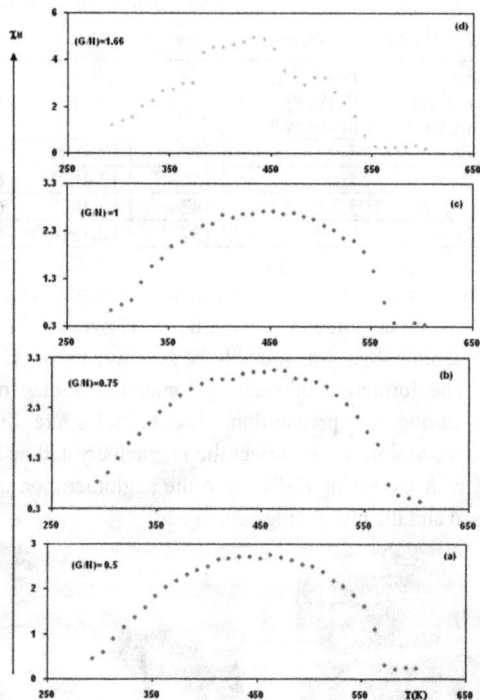

Fig. 5 a-d. Variation of (χ_M) with absolute temperature for $Gd_3Fe_5O_{12}$ at different (G/N) ratios.

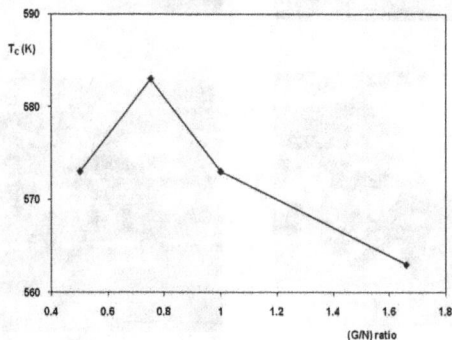

Fig. 5 e. The variation of the Curie Temperature with (G/N) ratio for the GdIG samples.

The temperature dependence of the molar magnetic susceptibility (χ_M) is illustrated in Fig. 5 a-d at magnetic field intensity of 1990 Oe from 300K to 700K. The data show that the same trend of χ_M is obtained irrespective of the (G/N) ratio, but with different values of χ_M and T_C. The behavior of χ_M is well known for all garnets [26] where a large broad hump is observed around 450 K, which shifted to lower temperatures with increasing the (G/N) ratio. This hump is characteristic for the iron garnet [26, 27]. The variation in Curie temperature with increasing (G/N) ratio is plotted in Fig. (5e). The decrease in the T_C with (G/N) ratio (except at 0.75) is associated with the weakening of the exchange interaction.

Gadolinium iron garnet ($Gd_3Fe_5O_{12}$–GdIG) is a total substitution of yttrium by gadolinium. The importance of this is the existence of a region in the magnetization versus temperature curve [13], between the compensation point, TCP, and the Curie temperature, T_C, in which the magnetization is approximately constant (dM/dT~0), which makes the difference for some particular microwave devices [14]. Note that the Gd magnetic moment is insensitive to the crystalline field and to the superexchange anisotropy due to the absence of net orbital angular momentum (L=0).

4. Conclusions

Gadolinium iron garnet (GdIG) was successfully synthesized in garnet cubic structure at annealing temperature of 1150°C by using flash autocombustion and glycine as fuel with different ratios. From these results, it was concluded that the lattice parameter decreased while the density increased with (G/N) ratio. This decrease was related to the completion of the reaction at lower temperature.

References

[1]   Li Guo,K. Huang,Y. Chen, G. Li, L.Yuan, W. Peng, H. Yuan, S.Feng, Mild hydrothermal synthesis and ferrimagnetism of $Pr_3Fe_5O_{12}$ and $Nd_3Fe_5O_{12}$ garnets, Journal of Solid State Chemistry, 184 (2011) 1048-1053. https://doi.org/10.1016/j.jssc.2011.03.010

[2]   X. Guo, Y. Chen, G.Wang, Y. Zhang, J.Ge, X. Tang, Freddy Ponchel, Denis Rémiens, Xianlin Dong, Growth and characterization of yttrium iron garnet films on Si substrates by Chemical Solution Deposition (CSD) technique, Journal of Alloys and Compounds, 671, (2016) 234-237. https://doi.org/10.1016/j.jallcom.2016.02.072

[3]   O. Opuchovic, A. Beganskiene, A. Kareiva, Sol–gel derived $Tb_3Fe_5O_{12}$ and $Y_3Fe_5O_{12}$ garnets: Synthesis, phase purity, micro-structure and improved design of

morphology, Journal of Alloys and Compounds, 647, (2015) 189-197.
https://doi.org/10.1016/j.jallcom.2015.05.169

[4]   A. Z. Arsad, N.B. Ibrahim, The effect of Ce doping on the structure, surface
morphology and magnetic properties of Dy doped-yttrium iron garnet films
prepared by a sol–gel method, Journal of Magnetism and Magnetic Materials, 410
(2016) 128-136. https://doi.org/10.1016/j.jmmm.2016.03.013

[5]   S. Song, D. Sheptyakov, M. Alexander. Korsunsky, Hai M. Duong, Li Lu, High Li
ion conductivity in a garnet-type solid electrolyte via unusual site occupation of
the doping Ca ions, Materials & Design, 93 (2016) 232-237.
https://doi.org/10.1016/j.matdes.2015.12.149

[6]   G. F. Dionne, Molecular-field coefficients of rare-earth iron garnets, Journal of
Applied Physics 47, (1976) 4220-4222. https://doi.org/10.1063/1.323204

[7]   G. Winkler, Magnetic Garnets, Friedr. Vieweg & Sohn, Braunschweig, Ch. 2
(1981).

[8]   A. H. Eschenfelder, Magnetic Bubble Technology, Springer, New York, (1981).
https://doi.org/10.1080/09500830500038092

[9]   T. Yamagishi, J. Awaka, Y. Kawashima, M. Uemura, S. Ebisu, S. Chikazawa,S.
Nagata,Ferrimagnetic order in the mixed garnet $(Y_{1-x}Gd_x)_3Fe_5O_{12}$, Philosophical
Magazine, 85 (2005) 1819-1833. https://doi.org/10.1080/09500830500038092

[10]   M. Uemura, T. Yamagishi, S. Ebisu, S. Chikazawa, and S. Nagata, A double peak
of the coercive force near the compensation temperature in the rare earth iron
garnets, , Philosophical Magazine, 88 (2008) 209-228.
https://doi.org/10.1080/14786430701805582

[11]   R. J. Joseyphus, A. Narayanasamy, A. K. Nigam, and R. Krishnan , Effect of
mechanical milling on the magnetic properties of garnets, Journal of Magnetism
and Magnetic Materials, 296, (2006) 57-64.
https://doi.org/10.1016/j.jmmm.2005.04.018

[12]   Ch. Liu and Z. John Zhang , Size-Dependent Superparamagnetic Properties of Mn
Spinel Ferrite Nanoparticles Synthesized from Reverse Micelles, Chemistry of
Materials, 13 (2001) 2092-2096. https://doi.org/10.1021/cm0009470

[13]   S. C. Zanatta, L. F. Co´tica, A. PaesanoJr., S. N. deMedeiros, J. B. M.daCunha, B.
Hallouche, Mechanosynthesis of Gadolinium Iron Garnet, Journal of American of
Ceramic Society, 88 (2005) 3316-3321. https://doi.org/10.1111/j.1551-
2916.2005.00598.x

[14] P.B.A.Fechine, F.M.M.Pereira, M.R.P.Santos, F.P.Filho, A.S.deMenezes, R.S.de Oliveira, .C. Go´es, L.P.Cardoso, A.S.B.Sombra, Microstructure and magneto-dielectric properties of ferrimagnetic composite $GdIG_X:YIG_{1-X}$ at radio and microwave frequencies, Journal of Physics and Chemistry of Solids, 70 (2009) 804-810. https://doi.org/10.1016/j.jpcs.2009.03.009

[15] S. Geller, Magnetic Interactions and Distribution of Ions in the Garnets Journal of Applied Physics, 31 (1960) S30-S37.

[16] A. Paesano, S. C. Zanata, S. N. De Mediros, L. F. Cotica, and J. B. M. Da Cunha, Mechanosyn thesis of YIG and GdIG: A Structural and Mössbauer Study, Hyperfine Interactions, 161 (2005) 211-220. https://doi.org/10.1007/s10751-005-9193-1

[17] R. D. Purohit, A.Chesnaud, A. Lachgar, O. Joubert, M.T.Caldes, Y. Piffard, L.Brohan, Development of New Oxygen Ion Conductors Based on Nd_4GeO_8 and Nd_3GaO_6, Chemistry of Materials, 17 (2005) 4479-4485. https://doi.org/10.1021/cm050537h

[18] R. D. Purohit, A. K. Tyagi, Auto-ignition synthesis of nanocrystalline $BaTi_4O_9$ powder, Journal of Materials Chemistry, 12 (2002) 312-316. https://doi.org/10.1039/b103461h

[19] S. Bhaduri, S.B. Bhaduri, E. Zhou, Auto ignition synthesis and consolidation of $Al_2O_3-ZrO_2$ nano/nano composite powders, Journal of Materials Research, 13 (1998)156-165. https://doi.org/10.1557/JMR.1998.0021

[20] J.Kingsley, K.Suresh, K.C.Patil, Combustion synthesis of fine-particle metal aluminates, Journal of Materials Science, 25(1990) 1305-1312. https://doi.org/10.1007/BF00585441

[21] H. K. Varma, P. Mukundan, K. G. K. Warrier, A. D. Damodaran, Flash combustion synthesis of cerium oxide, Journal of Materials Science Letters, 9 (4) (1990) 377-379. https://doi.org/10.1007/BF00721003

[22] B. D. Cullity, S. R. Stock, Elements of X- ray Diffraction, 3rd ed., Prentice- Hall, Englewood Cliffs, NJ, (2001).

[23] A. Mali and A. Ataie, Influence of the metal nitrates to citric acid molar ratio on the combustion process and phase constitution of barium hexaferrite particles prepared by sol–gel combustion method, Ceramics International, 30 (2004) 1979-1983. https://doi.org/10.1016/j.ceramint.2003.12.178

[24]   P. A. Lessiny, Mixed-cation oxide powders via polymeric precursors, Ceramic
        Bulletine, 68(5)(1989) 1002-1007.

[25]   X.Deng, Ch.Dong, EPCryst: a computer program for solving crystal structures
        from powder diffraction data, Journal of Applied Crystallography, 44 (2011) 230-
        237. https://doi.org/10.1107/S0021889810053835

[26]   M. A. Ahmed, Samiha T. Bishay and S. I. El-Dek, Conduction mechanism and
        magnetic behavior of dysprosium strontium iron garnet (DySrIG) nanocrystals,
        Materials Chemistry and Physics, 126 (2011)780-785.
        https://doi.org/10.1016/j.matchemphys.2010.12.044

[27]   M. A. Ahmed, Samiha T.Bishay, S.I.El-Dek, S.S.Solyman, Memory effect of
        nanoparticles $Dy_{2.8}Sr_{0.2}Fe_5O_{12}$ (DySrIG), Smart materials and structure, 21 (2012)
        045010-045015. https://doi.org/10.1088/0964-1726/21/4/045010

Chapter 4

Structural and Magnetic Properties of Vanadium Substituted SrM and Europium Substituted BaM Hexaferrites

Sami H. Mahmood[1,a], Ahmad M. Awadallah[1,b], Ibrahim Bsoul[2,c], Yazan Maswadeh[3,d]

[1] Physics Department, The University of Jordan, Amman 11942, Jordan

[2] Physics Department, Al al-Bayt University, Mafraq 13040, Jordan

[3] Physics Department, Central Michigan University, Mount Pleasant 48859, MI, USA

[a]s.mahmood@ju.edu.jo, [b]ahmadmoh@yahoo.com,[c]ibrahimbsoul@yahoo.com, [d]nawabra251@gmail.com

Abstract

$SrFe_{12-x}V_xO_{19}$, $x = 0.2$, 0.4 and $Ba_{0.8}Eu_{0.2}Fe_{12}O_{19}$ hexaferrites were prepared by high energy ball milling and sintering at 1200° C. XRD measurements revealed that the V-substituted SrM samples exhibited the coexistence of the pure SrM magnetic phase with nonmagnetic $Sr_3(VO_4)_2$ vanadate and α-Fe_2O_3 phase. Also, the Eu-substituted BaM hexaferrite revealed the formation of a pure BaM phase coexisting with α-Fe_2O_3 secondary phase, and Eu-garnet minor phase. The magnetic parameters of the substituted samples were found to be of potential importance for practical applications. The results of the study suggest methods for the preparation of high quality SrM hexaferrites, and hexaferrite/garnet composites.

Keywords

Hexaferrites, Garnet, Structural Properties, Coercivity, Magnetic Properties

Contents

1.   Introduction

Magnetic materials play a very important role in our life; magnets have been used in a wide range of industrial and technological applications such as data processing, electronic devices, telecommunication, automobile industry, loudspeakers, instrumentation, power generation, motors for a wide spectrum of applications, and so on [1, 2]. Hexaferrites belong to a special kind of magnetic oxides which demonstrated potential for various applications including permanent magnet and microwave applications. Three decades after their discovery in the 1950s, the annual production of M-type hexaferrite dominated the world market of permanent magnets due to cost effectiveness, easy production, corrosion resistance, and low eddy current losses [3]. The production and characterization of hexaferrites serves in both fields of scientific fundamental research and technological applications as demonstrated by the growth of market demand, and the exponential increase of the annual number of publications and registered patents [4]. For example, more than 50% of the magnetic materials produced annually around the world consist of M-type barium hexaferrite (BaM) [5].

Soft ferrites are usually used in transformers [6] and magnetic recording heads [7], multilayer chip inductors (MLCIs) [8, 9], soft magnets in electronic components such as power supplies [9-11], and in microwave high frequency devices [5, 12]. However, hard hexaferrites such as BaM have large magnetization and high coercive field, and is thus suitable for use in high-density magnetic recording media [13-19],permanent magnets [5, 10, 20], magneto-optics [21], and microwave absorption devices in GHz range [11, 15, 22]. Electromagnetic interference (EMI) could lead to severe interruption of the functioning of electronically controlled systems due to the electronic pollution produced by gigahertz (GHz) electronic telecommunication systems. For this reason, interest in producing and modifying electromagnetic wave absorbers and filters grew in the last few years. The M-type hexaferrites with planar magnetic anisotropy have better properties as electromagnetic wave absorbers in the GHz range compared to conventional Ni-Zn and Mn-Zn cubic ferrites [23, 24]. Cubic ferrites are widely used in various inductive devices with working frequencies below 100 MHz, but the problem with those ferrites arise when the working frequencies exceed 100 MHz [9, 25]. On the contrary, M-type hexaferrites have large dielectric and magnetic losses in the microwave frequency band [25]. In addition to their low production cost [26], low density, simple and easy fabrication process, and excellent chemical and thermal stability [11, 13, 15, 27], M-type hexaferrites have gained special attention in the field of materials research due to their interesting and attractive properties. These include, and are not limited to large saturation magnetization [15, 28], high coercive field with strong anisotropy along the c-axis [29, 30], high Curie temperature [11], very low electrical conductivity [13, 23], mechanical hardness [13], corrosion resistivity [11, 31], high microwave magnetic loss [22], moderate permittivity [32], and chemical compatibility with biological tissues [13, 33].

Based on their crystal structures and chemical compositions, barium-based hexaferrites are classified into six main types [34]. With the spinel formula (S) referring to $Me_2Fe_4O_8$ these types are:

- M-type: $[BaFe_{12}O_{19}] = [(BaO).6(Fe_2O_3)] = M$

- W-type: $[BaMe_2Fe_{16}O_{27}] = [(BaO)\cdot2(MeO)\cdot8(Fe_2O_3)] = M+S$

- X-type: $[Ba_2Me_2Fe_{28}O_{46}] = [2(BaO)\cdot2(MeO)\cdot14(Fe_2O_3)] = 2M+S = M+W$

- Y-type: $[Ba_2Me_2Fe_{12}O_{22}] = [2(BaO)\cdot2(MeO)\cdot6(Fe_2O_3)] = Y$

- Z-type: $[Ba_3Me_2Fe_{24}O_{41}] = [3(BaO)\cdot2(MeO)\cdot12(Fe_2O_3)] = M+Y$

- U-type: $[Ba_4Me_2Fe_{36}O_{60}] = [4(BaO)\cdot2(MeO)\cdot18(Fe_2O_3)] = 2M+Y$

In all of these ferrites, Me represents divalent metal ion and Ba can be replaced by Sr, Pb, or other elements [35]. The structure of BaM hexaferrite is hexagonal with $P6_3/mmc$ space group and lattice parameters $a = b = 5.89$Å, $c = 23.17$Å, $\alpha = \beta = 90°$, $\gamma = 120°$ [5]. The unit cell of BaM with magnetoplumbite-type structure includes two $BaFe_{12}O_{19}$ molecules as seen in Fig. 1 [36], each molecule arranged in two structural blocks: the hexagonal R block and the spinel S block [4, 36].

The unit cell is composed of the stacking sequence SRS^*R^*, where S^* and R^* blocks are S and R blocks rotated by 180° about the hexagonal c-axis. Within the S block, there are three interstitial sites between the oxygen layers which are occupied by the small metallic ions like Fe^{3+} ions [32]. One of these is the six-coordinated octahedral (2a) site occupied by spin-up iron ion, while the other two are four-coordinated tetrahedral ($4f_1$) sites occupied by spin-down iron ions. Also, within the R block there are three interstitial sites

115

between oxygen layers which are occupied by the small metallic ions, two six-fold coordinated octahedral ($4f_2$) sites occupied by spin-down iron ions, and one trigonal bi-pyramidal five-fold local symmetry (2b) site occupied by spin-up iron cation. Three interstitial octahedral (12k) sites are also available within each R-S interface layer, which are occupied by spin-up metallic ions.

Fig. 1. BaM unit cell.

Table. 1 shows the spin orientation of the Fe^{3+} at the different sites in the unit cell. According to Gorter's model of a system of collinear spin structure [37], the net magnetic moment per molecule is given by the sum of the magnetic moments of magnetic ions in the molecule. For BaM molecule ($BaFe_{12}O_{19}$), the S block contributes 2 spin down and 1 spin up moments, the R block contributes 2 spin down and 1 spin up moments, and both S-R interfaces contribute 6 spin up moments. Accordingly, the net magnetic moment per formula ($BaFe_{12}O_{19}$) can be calculated as follows.

$$m = (1\mu_{2a} - 2\mu_{4f1})_S + (6\mu_{12k})_{S-R} + (1\mu_{2b} - 2\mu_{4f2})_R = 4\mu \tag{1}$$

With a moment of $\mu = 5\mu_B$ per Fe^{3+} ion, the magnetic moment of $BaFe_{12}O_{19}$ is $20\mu_B$ per molecule. This corresponds to magnetization value of about 100 emu/g. This value was confirmed by measurements of the magnetization of SrM at low temperature where the thermal effects are minimized [38]. At room temperature, the magnetization for a typical BaM material was reduced to about 72 emu/g by thermal effects [5]. Also, typical coercive fields of 3 – 5 kOe were observe for BaM materials [5]. However, large variations in saturation magnetizations and coercive fields were reported, and were attributed to variations in particle size and morphology, as well as to different preparation methods [35]. Also, the substitution of Ba and/or Fe by other metals was found to influence the magnetization and coercivity significantly [35].

Table1. The spin orientation with its corresponding site for the unit cell.

Block	Sites	symmetry	Cations/site	Spin Orientation	Total spin magnetic moment
S	2a	Octahedral	1	Spin up (↑)	1 (↓)
	4f1	Tetrahedral	2	Spin down (↓)	
S-R	12k	Octahedral	3	Spin up (↑)	3 (↑)
R	2b	Bi-pyramidal	1	Spin up (↑)	1 (↓)
	4f2	Octahedral	2	Spin down (↓)	
R-S*	12k	Octahedral	3	Spin up (↑)	3 (↑)
S*	2a	Octahedral	1	Spin up (↑)	1 (↓)
	4f1	Tetrahedral	2	Spin down (↓)	
S*-R*	12k	Octahedral	3	Spin up (↑)	3 (↑)
R*	2b	Bi-pyramidal	1	Spin up (↑)	1 (↓)
	4f2	Octahedral	2	Spin down (↓)	
R*-S	12k	Octahedral	3	Spin up (↑)	3 (↑)

The effect of rare-earth ions substitution for the divalent metal ions was found to result in significant variations of the magnetic properties of M-type hexaferrites. Specifically, the substitution of Ba by Eu in $Ba_{0.75}Eu_{0.25}Fe_{12}O_{19}$ prepared sol–gel method was found to result in an increase in coercivity and a decrease in saturation magnetization [39]. Also, 10% substitution of Ba by La in BaM prepared by a reverse micro-emulsion route was found to lead to high magnetic properties, and enhancement of the microwave absorption properties for potential microwave absorption applications [40]. In addition, it was found that the saturation magnetization of $Sr_{1-x}La_xFe_{12-x}Co_xO_{19}$ improved upon increasing x up to 0.2, and the coercivity increased with increasing x up to 0.3 [41]. Further, La and Pr substitution for Ba in BaM powders prepared by auto-combustion was found to result in an increase in saturation magnetization and coercivity of the ferrite [42].

This work is a continuation of the recently reported study on the effect of vanadium substitution for iron on the structural and magnetic properties of BaM hexaferrites [18]. In addition, the study of the effect of RE substation was motivated by the reported influence of RE elements on the properties of M-type hexaferrites.

2.   Experimental procedure

Powder precursors of M-type hexaferrites were prepared by wet grinding using a high energy ball mill (Fritsch Pulverisette-7). $SrFe_{12-x}V_xO_{19}$ (with x = 0.2, 0.4) was prepared with high purity (> 98%) $SrCO_3$, Fe_2O_3, and V_2O_3. Required amounts of these materials were weighed carefully, and 5 g of the starting powder was transferred to each of the two zirconia cups of the ball-mill with 8 mL of acetone. Seven zirconia balls of approximately 10 g each were used for grinding, so that the powder-to-ball mass ratio was 1:14. The grinding was maintained for 16 hours with 250-rpm rotational speed, and the grinding was carried out in intervals of 10 min. each separated by a 5 min. pause period to avoid overheating. The dried powder was collected, and about 0.8 g discs (1.25 cm in diameter and ~ 1 mm thick) was prepared by pressing under a force of 40 kN. The discs were then sintered at a temperature of 1200° C for 2 h in a zirconium oxide crucible. Also, a $Ba_{0.8}Eu_{0.2}Fe_{12}O_{19}$ sample was prepared using the same procedure, except that the sintering process was carried out in an alumina crucible.

Structural details of the samples were obtained by analyzing the powder X-ray diffraction patterns recorded using XRD 7000-Shimadzu diffractometer with Cu-Kα radiation (λ = 1.5405 Å). The samples were scanned over the angular range $20° < 2\theta < 70°$ with 0.01° scanning step and speed of 0.5 deg/min. The grain morphology and grain size of the samples were examined with a Scanning Electron Microscope (SEM) system (FEI-Inspect F50/FEG). The magnetic properties of the samples were investigated using a vibrating sample magnetometer (VSM Micro Mag 3900, Princeton Measurements Corporation). Samples for VSM measurements were prepared by cutting small pieces from the sintered discs and gentle polishing to the desired needle-shape, to reduce the shape anisotropy effect. The magnetic data were recorded at room temperature under an applied magnetic field up to 10 kOe.

3.   Results and analysis

3.1   Structural results

The XRD patterns of $SrFe_{12-x}V_xO_{19}$ (x = 0.2, 0.4) samples are shown in Fig. 2. Rietveld refinement of the powder diffraction patterns indicated that each sample is composed of four structural phases: $SrFe_{12}O_{19}$, α-Fe_2O_3, $Sr_3(VO_4)_2$, and ZrO_2. The fitting was reliable as indicated by the low goodness of fit (χ^2 = 1.18 for the sample with x = 0.2, and 1.25 for the sample with x = 0.4). The appearance of the zirconia phase in the samples is an indication that the samples were contaminated as a result of the high sintering temperature in a zirconium oxide crucible, and this result is consistent with the appearance of this phase in similarly heat treated Cr_2Y samples [34]. The appearance of the α-Fe_2O_3 phase, on the other hand, resulted from the consumption of a fraction of Sr^{2+} ions commensurate with the stoichiometry of the $Sr_3(VO_4)_2$ vanadate phase, leaving the Fe:Sr molar ratio in the sample higher than the stoichiometric ratio of SrM phase. The reliability factors of Rietveld refinement and the phase proportions (in wt. %) are shown in Table 2.

The results of the refinement of the XRD pattern for the sample with $x = 0.2$ indicated that the SrM hexaferrite was the major phase, and that the iron oxide and zirconium oxide phases coexisted with appreciable weight fractions of the sample. The $Sr_3(VO_4)_2$ phase appeared with a relatively low weight fraction due to the small amount of vanadium in the sample. The effect of the vanadium substitution, however, is more effective in producing α-Fe_2O_3 phase, since the formation of one mole of vanadate (requiring 3 moles of Sr) should be accompanied with a surplus of 36 moles of Fe (18 moles of α-Fe_2O_3), since the Sr:Fe in the SrM is 1:12. The refinement results, therefore, indicated that the weight fraction of the α-Fe_2O_3 phase increased significantly in the sample with $x = 0.4$, which is due to the higher consumption of Sr by the vanadate phase, and the consequent depletion of Sr available for the formation of the SrM phase.

Phase contribution in the XRD peaks was labeled as S: M-type Strontium hexaferrite ($SrFe_{12}O_{19}$), α:Hematite (α-Fe_2O_3), V: Strontium Vanadium Oxide ($Sr_3(VO_4)_2$), Z: Zirconium (ZrO_2).

Table 2. Rietveld fitting reliability factors and fractions of phases (in wt. %) in V-substituted SrM hexaferrite samples.

Phase formula	R_B		R_F		JCPDS File #	Wt. %	
	$x = 0.2$	$x = 0.4$	$x = 0.2$	$x = 0.4$		$x = 0.2$	$x = 0.4$
$SrFe_{12}O_{19}$	4.38	3.22	3.11	2.51	00-024-1207	62.38	39.68
$Sr_3(VO_4)_2$	8.28	4.77	6.86	2.65	00-029-1318	3.17	3.99
α-Fe_2O_3	4.71	1.98	3.23	1.37	00-013-0534	23.58	43.81
ZrO_2	3.74	2.88	3.03	1.88	00-013-0307	10.87	12.52

The crystallite sizes of the different samples were calculated using the Stokes-Wilson relation [43].

$$D = \frac{\lambda}{\beta \cos\theta} \tag{2}$$

Here D is the crystallite size, λ is the wavelength of radiation (1.5406 Å), β is the integral breadth, and θ is the peak position. The integral breadth was corrected for instrumental broadening using Si standard sample. The crystallite size was determined from the (110) peak at $2\theta = 30.3°$, the (107) peak at $2\theta = 32.2°$, and the (114) peak at $2\theta = 34.1°$. Table 3 shows the crystallite size along the different crystallographic directions. These results indicate that the average crystallite size along the basal plane (the <110> direction) is larger than that along the c-axis, suggesting that the crystallites grow in platelet-like shapes. Further, higher concentrations of V substitution resulted in a significant reduction of the crystallite size, indicating poor crystallinity. This effect could be due to the hindrance of the growth of the SrM crystallite by the crystallization of the secondary phases in the sample.

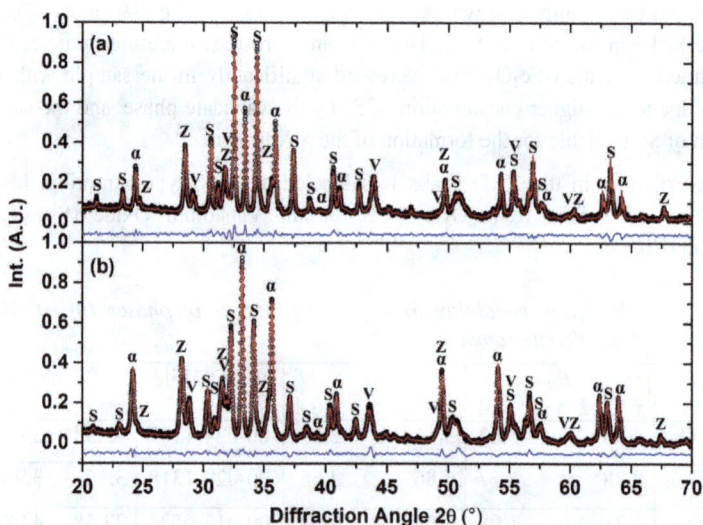

Fig. 2. XRD patterns for the samples $SrFe_{12-x}V_xO_{19}$: (a) x = 0.2, (b) x = 0.4.

Table 3. Crystallite size along different crystallographic directions for $SrFe_{12-x}V_xO_{19}$ (x = 0.2, 0.4) hexaferrites.

Vanadium content (x)	D (nm)		
	(1 1 0)	(1 0 7)	(1 1 4)
0.2	141	93	114
0.4	88	52	35

Fig. 3 shows the XRD pattern for the sample $Ba_{0.8}Eu_{0.2}Fe_{12}O_{19}$. Rietveld fitting of the diffraction pattern of this sample indicated that the experimental pattern is well fitted (χ^2 = 1.13) with a superposition of the patterns of $BaFe_{12}O_{19}$, α-Fe_2O_3, and the garnet phase $Eu_3Fe_5O_{12}$. The reliability factors and the fractions (in wt. %) of the phases derived from Rietveld refinement of the diffraction pattern are listed in Table 4.

Table 4. Fitting reliability factors and fractions of phases in the sample $Ba_{0.8}Eu_{0.2}Fe_{12}O_{19}$, determined by Rietveld refinement of the X-ray diffraction pattern.

Phase	R_B	R_F	JCPDS File #	Wt.%
$BaFe_{12}O_{19}$	2.47	2.52	00-027-1029	79.9
α-Fe_2O_3	3.26	3.13	00-013-0534	17.5
$Eu_3Fe_5O_{12}$	5.39	4.53	00-023-1069	2.60

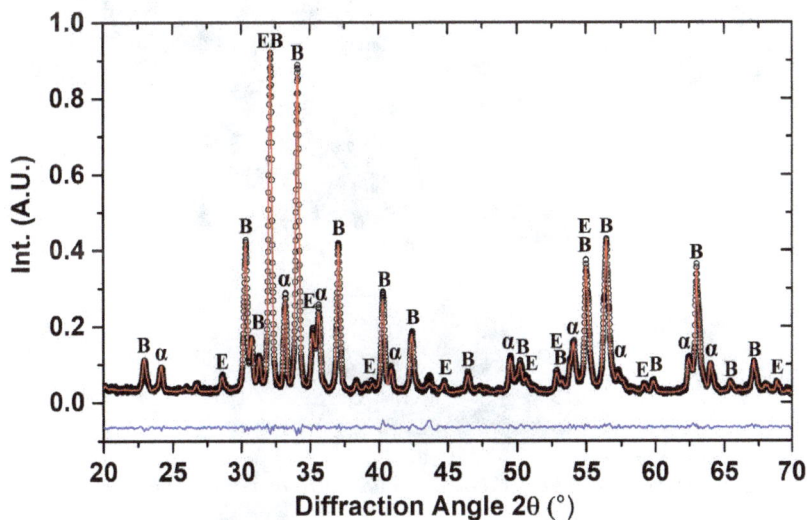

Fig. 3: XRD pattern of $Ba_{0.8}Eu_{0.2}Fe_{12}O_{19}$ sample. Phase contribution in the XRD peaks was labeled as B: M-type barium hexaferrite ($BaFe_{12}O_{19}$), α: Hematite (α-Fe_2O_3), E: Europium Iron Oxide ($Eu_3Fe_5O_{12}$).

These results indicated that the Eu-garnet phase is present with a very low fraction, compatible with the amount of Eu in the sample. According to the refinement results, the sample was phase separated according to the reaction scheme:

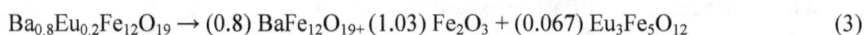

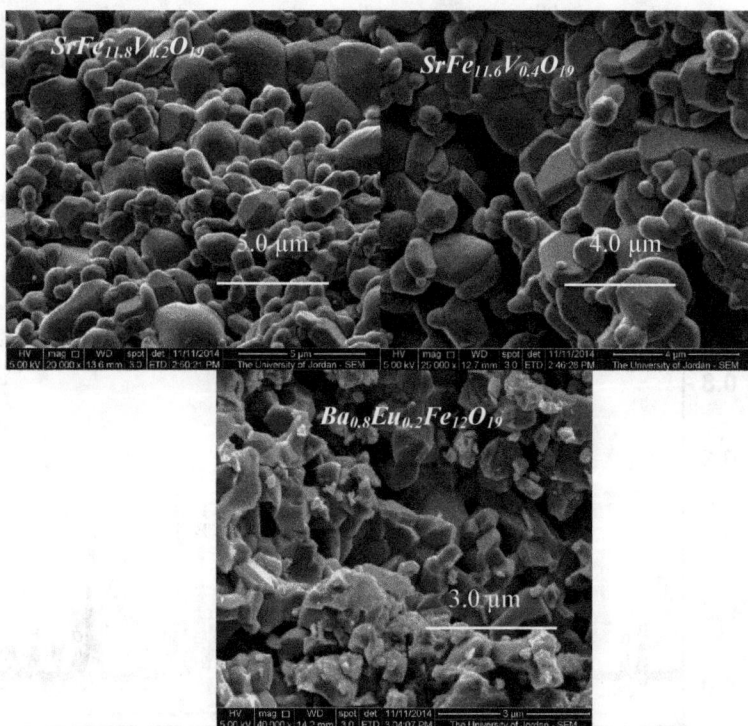

$$Ba_{0.8}Eu_{0.2}Fe_{12}O_{19} \rightarrow (0.8)\ BaFe_{12}O_{19+}\ (1.03)\ Fe_2O_3 + (0.067)\ Eu_3Fe_5O_{12} \qquad (3)$$

Fig. 4. SEM images of SrFe$_{12-x}$V$_x$O$_{19}$ (x = 0.2, 0.4) and Ba$_{1-x}$Eu$_x$Fe$_{12}$O$_{19}$ (x = 0.2) samples.

The crystallite size was calculated using Eq. 2. The average crystallite size along the <110> direction was found to be 34 nm, which is essentially the same as the value along the <107> direction. The crystallite size in the <114> direction, however, was found to be 50 nm. This indicates that the crystallites of BaM seem to grow into cuboidal shapes rather than platelet shapes. The relatively small crystallite size in this sample could also be associated with the hindrance of the growth of BaM crystallites by the crystallization of neighboring secondary phases.

3.2   SEM measurements

The SEM images of $SrFe_{12-x}V_xO_{19}$ (x = 0.2, 0.4) and $Ba_{1-x}Eu_xFe_{12}O_{19}$ (x = 0.2) samples are shown in Fig. 5. It can be seen that all samples show agglomerations of particles with generally wide particle size distributions for all samples. The particle size for the sample $SrFe_{11.8}V_{0.2}O_{19}$ ranged from about 400 nm to 3.0 μm, with appreciable fraction of the particles characterized by particle sizes below 1 μm.

On the other hand, the SEM image of the $SrFe_{11.6}V_{0.4}O_{19}$ sample shows particles ranging from 400 nm up to 3.6 μm in size, and general particle growth manifested by the reduction of the fraction of smaller particles in this sample. SEM image of the $Ba_{0.8}Eu_{0.2}Fe_{12}O_{19}$ sample shows particles with generally smaller sizes compared with the V-substituted SrM system. The particle size for this sample ranged from 200 nm to 1300 nm, with the majority of the particles characterized by particle sizes below 1 μm.

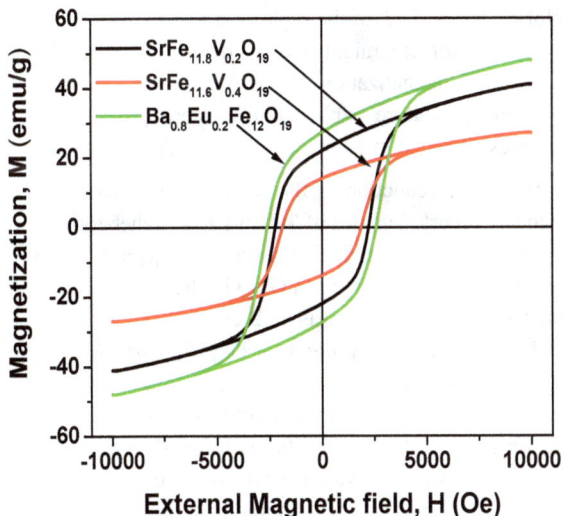

Fig. 5. Hysteresis loops for $SrFe_{12-x}V_xO_{19}$ (x = 0.2, 0.4) and $Ba_{0.8}Eu_{0.2}Fe_{12}O_{19}$ samples.

3.3 Magnetic measurements

The hysteresis loops of all samples were recorded at room temperature using VSM under an applied field up to 10 kOe, and are shown in Fig. 5. The remanence magnetization and the coercive fields were determined directly from the loops.

The hysteresis loops showed a behavior characteristic of hard magnetic material, indicated by the absence of magnetic saturation at the upper limit of the applied field.

Table 5. Saturation magnetization (M_s), remanence (M_r), squareness ratio ($M_{rs} = M_r/M_s$), coercive field (H_c), and anisotropy field (H_a) for the samples $SrFe_{12-x}V_xO_{19}$ ($x = 0.2, 0.4$) and $Ba_{0.8}Eu_{0.2}Fe_{12}O_{19}$.

Substitution (x)	M_s (emu/g)	M_r (emu/g)	M_{rs} (emu/g)	H_c (kOe)	H_a (kOe)
SrM-0.2 V	45.2	21.9	0.49	2.22	11.6
SrM-0.4 V	29.4	13.9	0.47	1.87	10.9
BaM-0.2 Eu	53.1	27.2	0.51	2.59	11.9

Accordingly, the law of approach to saturation was used to obtain the saturation magnetization. A plot of Ms vs. $1/H^2$ in the high field region (8.6 kOe $<H<$ 10 kOe) for the different samples gave perfect straight lines as demonstrated by Fig. 6. From these straight lines, the saturation magnetization M_s, as well as the anisotropy field H_a were determined following the procedure explained elsewhere [16-19, 30]. The magnetic parameters of the samples are listed in Table 5.

The magnetic data revealed a reduction of the saturation magnetization of all samples with respect to previously reported values of 70 emu/g or higher for pure BaM and SrM compounds [5, 35, 44, 45]. This reduction is a consequence of the formation of nonmagnetic phases (α-Fe_2O_3, vanadate- $Sr_3(VO_4)_2$) in the samples. The value of M_s for the V-substituted SrM sample with $x = 0.2$ is 45.2 emu/g, which is lower than the value of 56.8 emu/g for BaM with 0.2 V-substitution for Fe [18]. Considering that the wt. % of the hexaferrite magnetic phase in this sample is 62.38%, the saturation magnetization of the hexaferrite phase (normalized to its weight fraction) would be 72.5 emu/g. Also, the value of 29.4 emu/g for the sample with $x = 0.4$ is lower than that of 38 emu/g for the similarly substituted BaM [18]. However, the wt. % of the hexaferrite phase in this sample is only 39.68 %, implying that the saturation magnetization of the hexaferrite phase is 74.1 emu/g. The normalized saturation magnetizations are in agreement with the highest values reported for pure SrM hexaferrites [5]. Accordingly, this synthesis route can be adopted to produce high quality SrM magnets by following the procedure recently reported for the production of BaM hexaferrites [18, 46], namely, adding an extra amount of Sr enough to eliminate the occurrence of the α-Fe$_2$O$_3$ phase, avoiding sintering the powder mixture in the zirconium oxide crucible, and washing the sintered powder with HCl to etch out the vanadate nonmagnetic phase. The coercivities of these samples of 2.22 kOe and 1.87 kOe, respectively, are lower than those of similarly substituted BaM [18]. The reduced coercivity could be due to larger particle size in substituted SrM samples. The magnetic properties of these samples could be of potential importance for high density magnetic recording applications [47].

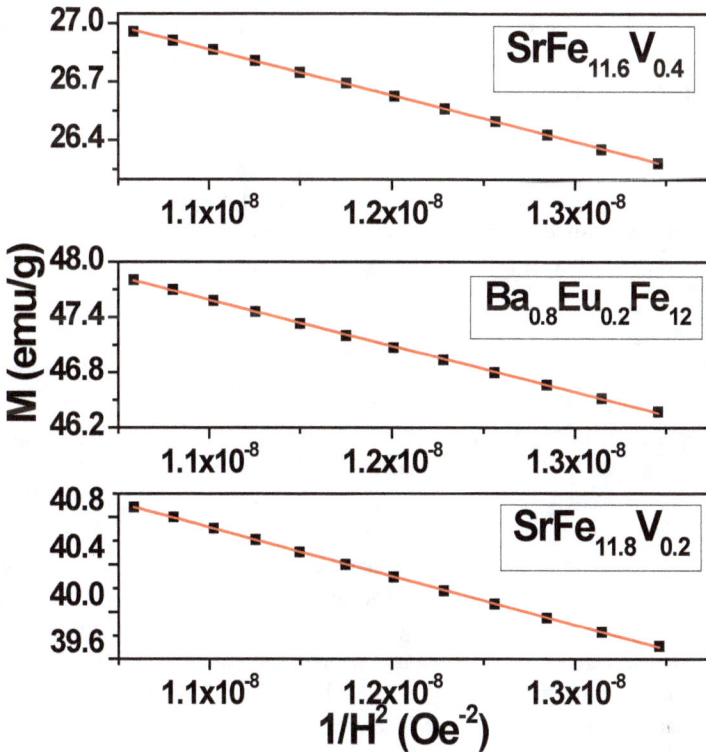

Fig. 6. Magnetization versus $1/H^2$ for SrFe$_{12-x}$V$_x$O$_{19}$ (x = 0.2, 0.4) and Ba$_{0.8}$Eu$_{0.2}$Fe$_{12}$O$_{19}$ samples.

The sample $Ba_{0.8}Eu_{0.2}Fe_{12}O_{19}$ also exhibited a saturation magnetization of 53.1 emu/g, and coercivity of 2.59 kOe. The saturation magnetization is only slightly higher than that reported by Khademi et al. [39] for similarly doped BaM samples prepared by sol-gel method, who found these materials of potential importance for microwave applications in the GHz region. Considering the weight fraction of the BaM phase of 79.9%, and ignoring the small magnetic contribution of the 2.6 wt. % of the garnet phase, the saturation magnetization of the BaM phase, normalized to its weight fraction in the sample, would be 66.5 emu/g. This value is close to the highest reported experimental values for pure BaM hexaferrites [35]. The results of the present work suggests a method for the preparation of hexaferrite/garnet composites with desired composition. This would involve the choice of the Eu amount to provide the required fraction of the garnet phase, adding the amount of Fe necessary for that phase, and then adding stoichiometric amounts of Ba and Fe necessary for the BaM phase. The product composite consisting of the self-biased BaM phase [32] together with the soft garnets could therefore be tuned for microwave applications in the desired microwave band.

The magnetocrystalline anisotropy field (H_a) for all investigated samples remained close to the reported value of ~ 12 kOe for pure and substituted BaM [30, 48]. This is due to the fact that the anisotropy field is characteristic of a pure magnetic phase, and is not influenced by the coexisting nonmagnetic phases. Further, the values of M_{rs}~ 0.5 are consistent with the existence of particles consisting of single magnetic domains of size $\leq$ 1 μm [5].

Conclusions

The partial substitution of Fe by V in SrM hexaferrite resulted in phase separation and the consequent appearance of vanadate and α-Fe_2O_3 nonmagnetic phases coexisting with the magnetic hexaferrite phase. This phase separation resulted in a reduction of the saturation magnetization and coercivity with respect to pure SrM hexaferrite. The sample with 0.2 V-substitution exhibited magnetic properties which could be of importance for high density magnetic recording applications. On the other hand, partial substitution of Ba by Eu in BaM hexaferrite was not successfully achieved. Instead, the Eu was involved in forming Eu-garnet phase, and the remaining precursor powders formed pure BaM and α-Fe_2O_3. As a consequence, the saturation magnetization of the sample decreased. However, the saturation magnetization of the sample remained high enough for practical applications, and the reduction of the coercivity makes this sample of potential importance for high density magnetic recording.

Acknowledgements

This work was supported by the Deanship of scientific Research and Quality Assurance at The University of Jordan.

References

[1]   Ü. Özgür, Y. Alivov, H. Morkoç, Microwave ferrites, part 1: fundamental properties, Journal of Materials Science: Materials in Electronics, 20 (2009) 789-834. https://doi.org/10.1007/s10854-009-9923-2

[2]   S. H. Mahmood, I. Abu-Aljarayesh, Hexaferrite Permanent Magnetic Materials, Materials Research Forum LLC, Millersville, PA, USA, 2016. https://doi.org/10.21741/9781945291074

[3]   S.H. Mahmood, High Performance Permanent Magnets, in: S.H. Mahmood, I. Abu-Aljarayesh (Eds.) Hexaferrite Permanent Magnetic Materials, Materials Research Forum LLC, Millersville, PA, 2016, pp. 47-73. https://doi.org/10.21741/9781945291074

[4]   S.H. Mahmood, Properties and Synthesis of Hexaferrites, in: S.H. Mahmood, I. Abu-Aljarayesh (Eds.) Hexaferrite Permanent Magnetic Materials, Materials Research Forum LLC, Millersville, PA, 2016, pp. 74-110. https://doi.org/10.21741/9781945291074

[5]   R.C. Pullar, Hexagonal ferrites: A review of the synthesis, properties and applications of hexaferrite ceramics, Progress in Materials Science, 57 (2012) 1191-1334. https://doi.org/10.1016/j.pmatsci.2012.04.001

[6]   O. Masala, D. Hoffman, N. Sundaram, K. Page, T. Proffen, G. Lawes, R. Seshadri, Preparation of magnetic spinel ferrite core/shell nanoparticles: Soft ferrites on hard ferrites and vice versa, Solid state sciences, 8 (2006) 1015-1022. https://doi.org/10.1016/j.solidstatesciences.2006.04.014

[7]   J. Kulikowski, A. Leśniewski, properties of Ni-Zn ferrites for magnetic heads: technical possibilities and limitations, Journal of Magnetism and Magnetic Materials, 19 (1980) 117-119. https://doi.org/10.1016/0304-8853(80)90569-7

[8]   S. Mahmood, F. Jaradat, A.-F. Lehlooh, A. Hammoudeh, Structural properties and hyperfine interactions in Co–Zn Y-type hexaferrites prepared by sol–gel method, Ceramics International, 40 (2014) 5231-5236. https://doi.org/10.1016/j.ceramint.2013.10.092

[9]   W. Zhang, Y. Bai, X. Han, L. Wang, X. Lu, L. Qiao, Magnetic properties of Co–
        Ti substituted barium hexaferrite, Journal of Alloys and Compounds, 546 (2013)
        234-238. https://doi.org/10.1016/j.jallcom.2012.08.029

[10]   J. Smit, H.P.J. Wijn, Ferrites: physical properties of ferrimagnetic oxides in
        relation to their technical applications, Wiley1959.

[11]   S. Dursun, R. Topkaya, N. Akdoğan, S. Alkoy, Comparison of the structural and
        magnetic properties of submicron barium hexaferrite powders prepared by molten
        salt and solid state calcination routes, Ceramics International, 38 (2012) 3801-
        3806. https://doi.org/10.1016/j.ceramint.2012.01.028

[12]   H.-S. Cho, S.-S. Kim, M-hexaferrites with planar magnetic anisotropy and their
        application to high-frequency microwave absorbers, Magnetics, IEEE
        Transactions on, 35 (1999) 3151-3153.

[13]   K. Martirosyan, E. Galstyan, S. Hossain, Y.-J. Wang, D. Litvinov, Barium
        hexaferrite nanoparticles: synthesis and magnetic properties, Materials Science
        and Engineering: B, 176 (2011) 8-13. https://doi.org/10.1016/j.mseb.2010.08.005

[14]   U. Topal, H. Ozkan, H. Sozeri, Synthesis and characterization of nanocrystalline
        $BaFe_{12}O_{19}$ obtained at 850 C by using ammonium nitrate melt, Journal of
        magnetism and magnetic materials, 284 (2004) 416-422.
        https://doi.org/10.1016/j.jmmm.2004.07.009

[15]   J. Qiu, L. Lan, H. Zhang, M. Gu, Microwave absorption properties of
        nanocomposite films of $BaFe_{12}O_{19}$ and TiO_2 prepared by sol–gel method,
        Materials Science and Engineering: B, 133 (2006) 191-194.
        https://doi.org/10.1016/j.mseb.2006.06.049

[16]   S.H. Mahmood, A. Awadallah, Y. Maswadeh, I. Bsoul, Structural and magnetic
        properties of Cu-V substituted M-type barium hexaferrites, IOP Conference
        Series: Materials Science and Engineering, IOP Publishing, 2015, pp. 012008.
        https://doi.org/10.1088/1757-899X/92/1/012008

[17]   A. Awadallah, S.H. Mahmood, Y. Maswadeh, I. Bsoul, M. Awawdeh, Q.I.
        Mohaidat, H. Juwhari, Structural, magnetic, and Mossbauer spectroscopy of Cu
        substituted M-type hexaferrites, Materials Research Bulletin, 74 (2016) 192-201.
        https://doi.org/10.1016/j.materresbull.2015.10.034

[18]   A. Awadallah, S.H. Mahmood, Y. Maswadeh, I. Bsoul, A. Aloqaily, Structural and
        magnetic properties of Vanadium Doped M-Type Barium Hexaferrite ($BaFe_{12-}$

$_xV_xO_{19}$), IOP Conference Series: Materials Science and Engineering, IOP Publishing, 2015, pp. 012006. https://doi.org/10.1088/1757-899X/92/1/012006

[19] S.H. Mahmood, A.A. Ghanem, I. Bsoul, A. Awadallah, Y. Maswadeh, Structural and magnetic properties of $BaFe_{12-2x}Cu_xMn_xO_{19}$ hexaferrites, Materials Research Express, 4 (2017) 036105. https://doi.org/10.1088/2053-1591/aa646c

[20] R.M. Almeida, W. Paraguassu, D.S. Pires, R.R. Correa, C.W. de Araujo Paschoal, Impedance spectroscopy analysis of $BaFe_{12}O_{19}$ M-type hexaferrite obtained by ceramic method, Ceramics International, 35 (2009) 2443-2447. https://doi.org/10.1016/j.ceramint.2009.02.020

[21] N. Dishovski, A. Petkov, I. Nedkov, I. Razkazov, Hexaferrite contribution to microwave absorbers characteristics, Magnetics, IEEE Transactions on, 30 (1994) 969-971.

[22] A. Ghasemi, A. Morisako, Static and high frequency magnetic properties of Mn– Co–Zr substituted Ba-ferrite, Journal of Alloys and Compounds, 456 (2008) 485-491. https://doi.org/10.1016/j.jallcom.2007.02.101

[23] A. Ghasemi, A. Hossienpour, A. Morisako, X. Liu, A. Ashrafizadeh, Investigation of the microwave absorptive behavior of doped barium ferrites, Materials & Design, 29 (2008) 112-117. https://doi.org/10.1016/j.matdes.2006.11.019

[24] S. Sugimoto, K. Okayama, S.-i. Kondo, H. Ota, M. Kimura, Y. Yoshida, H. Nakamura, D. Book, T. Kagotani, M. Homma, Barium M-type ferrite as an electromagnetic microwave absorber in the GHz range, Mater. Trans. JIM, 39 (1998) 1080-1083. https://doi.org/10.2320/matertrans1989.39.1080

[25] S. Sugimoto, K. Haga, T. Kagotani, K. Inomata, Microwave absorption properties of Ba M-type ferrite prepared by a modified coprecipitation method, Journal of magnetism and magnetic materials, 290 (2005) 1188-1191. https://doi.org/10.1016/j.jmmm.2004.11.381

[26] V.V. Soman, V. Nanoti, D. Kulkarni, Dielectric and magnetic properties of Mg–Ti substituted barium hexaferrite, Ceramics International, 39 (2013) 5713-5723. https://doi.org/10.1016/j.ceramint.2012.12.089

[27] I. Bsoul, S.H. Mahmood, A.F. Lehlooh, Structural and magnetic properties of $BaFe_{12-2x}Ti_xRu_xO_{19}$, Journal of Alloys and Compounds, 498 (2010) 157-161. https://doi.org/10.1016/j.jallcom.2010.03.142

[28] I. Bsoul, S.H. Mahmood, Magnetic and structural properties of $BaFe_{12-x}Ga_xO_{19}$ nanoparticles, Journal of Alloys and Compounds, 489 (2010) 110-114. https://doi.org/10.1016/j.jallcom.2009.09.024

[29] I. Ali, M. Islam, M. Awan, M. Ahmad, M.N. Ashiq, S. Naseem, Effect of Tb^{3+} substitution on the structural and magnetic properties of M-type hexaferrites synthesized by sol–gel auto-combustion technique, Journal of Alloys and Compounds, 550 (2013) 564-572. https://doi.org/10.1016/j.jallcom.2012.10.121

[30] S.H. Mahmood, G.H. Dushaq, I. Bsoul, M. Awawdeh, H.K. Juwhari, B.I. Lahlouh, M.A. AlDamen, Magnetic Properties and Hyperfine Interactions in M-Type $BaFe_{12-2x}Mo_xZn_xO_{19}$ Hexaferrites, Journal of Applied Mathematics and Physics, 2 (2014) 77-87. https://doi.org/10.4236/jamp.2014.25011

[31] I. Bsoul, S.H. Mahmood, A.F. Lehlooh, A. Al-Jamel, Structural and magnetic properties of $SrFe_{12-2\,x}Ti_xRu_xO_{19}$, Journal of Alloys and Compounds, 551 (2013) 490-495. https://doi.org/10.1016/j.jallcom.2012.11.062

[32] V.G. Harris, A. Geiler, Y. Chen, S.D. Yoon, M. Wu, A. Yang, Z. Chen, P. He, P.V. Parimi, X. Zuo, Recent advances in processing and applications of microwave ferrites, Journal of Magnetism and Magnetic Materials, 321 (2009) 2035-2047. https://doi.org/10.1016/j.jmmm.2009.01.004

[33] Q.A. Pankhurst, J. Connolly, S. Jones, J. Dobson, Applications of magnetic nanoparticles in biomedicine, Journal of physics D: Applied physics, 36 (2003) R167. https://doi.org/10.1088/0022-3727/36/13/201

[34] S.H. Mahmood, M.D. Zaqsaw, O.E. Mohsen, A. Awadallah, I. Bsoul, M. Awawdeh, Q.I. Mohaidat, Modification of the magnetic properties of Co_2Y hexaferrites by divalent and trivalent metal substitutions, Solid State Phenomena, 241 (2016) 93-125. https://doi.org/10.4028/www.scientific.net/SSP.241.93

[35] S.H. Mahmood, Ferrites with High Magnetic Parameters, in: S.H. Mahmood, I. Abu-Aljarayesh (Eds.) Hexaferrite Permanent Magnetic Materials, Materials Research Forum LLC, Millersville, PA, 2016, pp. 111-152. https://doi.org/10.21741/9781945291074

[36] Y. Maswadeh, S.H. Mahmood, A. Awadallah, A.N. Aloqaily, Synthesis and structural characterization of nonstoichiometric barium hexaferrite materials with Fe: Ba ratio of 11.5–16.16, IOP Conference Series: Materials Science and Engineering, IOP Publishing, 2015, pp. 012019.

[37] E. Gorter, Saturation magnetization of some ferrimagnetic oxides with hexagonal crystal structures, Proceedings of the IEE-Part B: Radio and Electronic Engineering, 104 (1957) 255-260. https://doi.org/10.1049/pi-b-1.1957.0042

[38] A. Alsmadi, I. Bsoul, S. Mahmood, G. Alnawashi, F. Al-Dweri, Y. Maswadeh, U. Welp, Magnetic study of M-type Ru-Ti doped strontium hexaferrite nanocrystalline particles, Journal of Alloys and Compounds, 648 (2015) 419-427. https://doi.org/10.1016/j.jallcom.2015.06.274

[39] F. Khademi, A. Poorbafrani, P. Kameli, H. Salamati, Structural, magnetic and microwave properties of Eu-doped barium hexaferrite powders, Journal of superconductivity and novel magnetism, 25 (2012) 525-531. https://doi.org/10.1007/s10948-011-1323-1

[40] X. Gao, Y. Du, X. Liu, P. Xu, X. Han, Synthesis and characterization of Co–Sn substituted barium ferrite particles by a reverse microemulsion technique, Materials Research Bulletin, 46 (2011) 643-648. https://doi.org/10.1016/j.materresbull.2011.02.002

[41] L. Peng, L. Li, R. Wang, Y. Hu, X. Tu, X. Zhong, Microwave sintered $Sr_{1-x}La_xFe_{12-x}Co_xO_{19}$ ($x = $ 0–0.5) ferrites for use in low temperature co-fired ceramics technology, Journal of Alloys and Compounds, 656 (2016) 290-294. https://doi.org/10.1016/j.jallcom.2015.08.263

[42] S. Ounnunkad, Improving magnetic properties of barium hexaferrites by La or Pr substitution, Solid State Communications, 138 (2006) 472-475. https://doi.org/10.1016/j.ssc.2006.03.020

[43] B.E. Warren, X-ray Diffraction, Addison-Wesley, Reading, Massachsetts, 1969.

[44] G.H. Dushaq, S.H. Mahmood, I. Bsoul, H.K. Juwhari, B. Lahlouh, M.A. AlDamen, Effects of molybdenum concentration and valence state on the structural and magnetic properties of $BaFe_{11.6}Mo_xZn_{0.4-x}O_{19}$ hexaferrites, Acta Metallurgica Sinica (English Letters), 26 (2013) 509-516. https://doi.org/10.1007/s40195-013-0075-2

[45] S.H. Mahmood, I. Bsoul, Hopkinson peak and superparamagnetic effects in $BaFe_{12-x}Ga_xO_{19}$ nanoparticles, EPJ Web of Conferences, 29 (2012) 00039.

[46] U. Topal, Towards Further Improvements of the Magnetization Parameters of B_2O_3-Doped $BaFe_{12}O_{19}$ Particles: Etching with Hydrochloric Acid, Journal of superconductivity and novel magnetism, 25 (2012) 1485-1488. https://doi.org/10.1007/s10948-012-1486-4

[47]   I. Abu-Aljarayesh, Magnetic Recording, in: S.H. Mahmood, I. Abu-Aljarayesh (Eds.) Hexaferrite Permanent Magnetic Materials, Materials Research Forum LLC, Millersville, PA, 2016, pp. 166-181.

[48]   M. Awawdeh, I. Bsoul, S.H. Mahmood, Magnetic properties and Mössbauer spectroscopy on Ga, Al, and Cr substituted hexaferrites, Journal of Alloys and Compounds, 585 (2014) 465-473. https://doi.org/10.1016/j.jallcom.2013.09.174

Chapter 5

Synthesis and Investigation of Electrical and Magnetic Properties of Cobalt Substituted Lithium Nano Ferrites

G. Aravind[1,a], M. Raghasudha[2,b], D. Ravinder[3,c], P. Veerasomaiah[2,d]

[1]Methodist College of Engineering and Technology, Osmania University, Hyderabad-500 007, Telangana, India

[2]Department of Chemistry, University College of Science, Osmania University, Hyderabad-500 007, Telangana, India

[3]Department of Physics, University College of Science, Osmania University, Hyderabad-500 007, Telangana, India

[a]gunthaaravind@gmail.com, [b]raghasudha_m@yahoo.co.in, [c]ravindergupta28@rediffmail.com,[d]vs_puppala@rediffmail.com

Abstract

The chapter deals with the investigation of electrical and magnetic properties of $Li_{0.5-0.5x}Co_xFe_{2.5-0.5x}O_4$ (x = 0.0 to 1.0) synthesized by a citrate-gel auto combustion method. Cobalt substitution was reported to transform the soft pure lithium ferrite into magnetically hard cobalt substituted lithium ferrite. DC resistivity results show the semiconducting behavior. Dielectric constant and dielectric loss tangent of all prepared samples found to decrease with increasing the frequency. Magnetization results indicated that Li-Co ferrites (x = 0.8, x = 1.0) show superparamagnetic behavior and hence they can be useful in biomedical applications for targeted drug delivery and magnetic resonance imaging (MRI).

Keywords

Nano Ferrites, XRD, SEM, Electrical Properties, Magnetic Properties

Contents

1.   Introduction

During the last few decades, nano-materials have attracted enormous interest in the research community worldwide. These materials have novel, interesting and important properties due to their extremely small size, and they have the potential for wide-ranging applications [1]. Nano materials can be used to improve the properties of existing materials and create novel materials with interesting properties [2]. Among the nano materials, spinel nano ferrites show extraordinary electrical and magnetic properties depending on the composition. The field of ferrites is well developed, but due to its potential applications in various fields, and interesting physics involved in it, even after several decades of its first artificial synthesis, scientists and researchers are still interested in various types of ferrites with different preparation techniques. Ferrimagnetic materials (ferrites) have a high magnetic permeability which is used to store a stronger magnetic field than iron. Reddy *et al.* [3] studied the electrical properties of Li-Ni ferrites and found that the conductivity and dielectric constant of the prepared samples decreased with decreasing concentration of ferrous ions in the ferrites. The squareness and remanence ratio of lithium ferrites were found to increase with increasing amount of Ni [4]. Rajesh Cheruku *et al.* [5] prepared Bi and Pb doped lithium ferrites by solution combustion method, and from the VSM and ESR data, they confirmed that the synthesized nano particles show super-paramagnetic nature. S.T. Assar *et al.* [6] studied the dielectric behavior of lithium doped Co-Ni ferrites prepared by citrate precursor method and found that classical barrier hopping model is the most probable mechanism in the samples under investigation.

The properties of ferrites are deeply influenced by several factors like method of synthesis, grain size, amount of additives, microstructure of the material, pH value, sintering temperature/time and atmosphere etc. [7]. Among spinel ferrites, lithium ferrites are of special interest due to their potential applications in various fields. The use of lithium ferrites, particularly in microwave devices and batteries, is restricted due to the difficulties experienced in sintering the material at the high temperatures employed to achieve high densities ($\sim$ X-ray densities) in stoichiometric form. The irreversible loss of lithium and oxygen during sintering was the main reason that made lithium ferrites technologically difficult to synthesize [8].

In order to avoid the high temperature sintering, different wet chemical techniques were developed such as chemical co-precipitation [9], reverse micelle [10], micro emulsion method [11], and sol-gel technique [12]. In the sol-gel method, metal alkoxides are dissolved in an organic solvent like benzene, toluene etc., to get a sol, which is further decomposed to get the metal oxides. But, most of the metal alkoxides, whose valence is less than four, are not soluble in organic solvents. This sol-gel technique has been modified with the use of metal nitrates, which undergo self combustion when the gel is dried, producing the metal oxides. After thorough survey, the authors are interested in synthesizing cobalt substituted lithium nano ferrites and studying their electrical and magnetic properties. The present article reports the synthesis and study of electrical and magnetic properties of cobalt substituted lithium nano ferrites with the chemical composition $[Li_{0.5}Fe_{0.5}]_{1-x}Co_x Fe_2O_4$ (x = 0.0, 0.2, 0.4, 0.6, 0.8, 1.0).

2. Experimental procedure

Lithium Nitrate - Li(NO$_3$), (> 99.0%, Sigma-Aldrich), cobalt nitrate- Co(NO$_3$)$_2$·6H$_2$O ($\geq$ 98.0%, Sigma-Aldrich), , ferric nitrate-Fe(NO$_3$)$_3$·9H$_2$O ($\geq$ 98.0%, Sigma-Aldrich), citric acid- C$_6$H$_8$O$_7$H$_2$O (99%, Sigma-Aldrich) and ammonia-NH$_3$ (anhydrous 99.98% Aldrich) were used as starting materials for the synthesis of [Li$_{0.5}$Fe$_{0.5}$]$_{1-x}$Co$_x$Fe$_2$O$_4$ (x = 0.0, 0.2, 0.4, 0.6, 0.8, 1.0). Calculated quantities of metal nitrates of a particular nano ferrite composition along with citric acid were dissolved in a minimum quantity of double distilled water. The mixture was thoroughly stirred using a magnetic stirrer to get a homogeneous solution.

Ammonia solution was added to this mixture to adjust the pH to 7. Then the mixed solution was heated to about 80 °C with uniform stirring on a hot plate to obtain highly viscous gel called citrate precursor. When the volume of the solution reduced to one fourth of its initial volume, the temperature was increased to 200 °C. Finally, all water molecules were removed from the mixture, and then the viscous gel started burning automatically which gives burnt ash. The sudden combustion causes the gathering of an organic complex together and accelerates the solid-solid reaction resulting in the formation of nano particles at lower firing temperature [13]. The burnt powder was thoroughly ground using agate mortar-pestle and was calcined in air in a muffle furnace at 500 °C for 4 h. The resultant powder is the required nano crystalline ferrite sample which is ready to characterize.

As prepared ferrites were characterized by various experimental techniques like X-ray Diffraction analysis, Scanning Electron Microscopy (SEM), and Energy Dispersive Spectroscopy (EDS). Electrical properties such as DC electrical resistivity, dielectric properties of the prepared nano ferrites were studied by various experimental techniques. DC resistivity of the prepared samples was measured by using the two probe method in the temperature range of 200 – 600 °C. Dielectric parameters of the samples were measured by using a LCR meter up to the temperature 450° C in the frequency range of 20 Hz – 5 MHz. Room temperature and low temperature magnetization measurements of the Li-Co nano ferrites were studied by using a vibrating sample magnetometer.

3. Results and discussion

3.1 XRD analysis

The structural study is essential for optimizing the properties needed for various applications. The phase identification and structural parameter determination were examined on an X-ray diffractometer ((Philips X-ray diffractometer model 3710). The X-ray diffraction patterns of the $[Li_{0.5}Fe_{0.5}]_{1-x}Co_xFe_2O_4$ (where x = 0.0 to 1.0 with step of 0.2) ferrite samples, sintered at 500° C for 4 h are shown in Fig. 1.

Analyzing the X-ray diffraction patterns one can observe that the positions of the peaks comply with the reported values [14-15]. From Fig.1, it can be seen that the position of the peaks in XRD pattern of all prepared samples are coincident, which indicate that there were no distinct differences in phase structure of Li-Co nano crystalline ferrites with different cobalt composition.

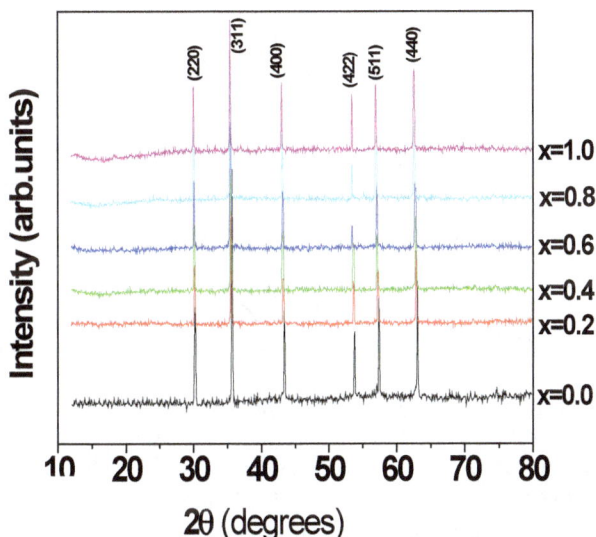

Fig. 1. XRD patterns of $[Li_{0.5}Fe_{0.5}]_{1-x}Co_xFe_2O_4$ (x = 0.0 to 1.0) ferrite samples

On increasing the cobalt compositions, reflection peaks become sharp and reflection intensities of the pattern increase which reveal that cobalt composition is useful in the crystallization of the nano crystalline ferrites and promote the grain growth. It was observed that cobalt ions were partly replaced by iron and lithium ions in the lattice and then increased the lattice constant.

The observed (220), (311), (400), (422), (511) and (440) reflections confirmed formation of spinel structure of the prepared samples. This indicates that the synthesized ferrite compositions are of single phase cubic spinel since no ambiguous reflections other than those of the spinel structure are observed and also demonstrates the homogeneity of the prepared nano crystalline ferrite samples.

The X-ray diffraction analysis of the prepared Li-Co nano crystalline ferrite samples helps to estimate the crystallographic parameters. The lattice constant (a), crystalline size (D_{xrd}), X-ray density and experimental density are reported in table 1. The average crystallite size of the samples was calculated from the Debye-Scherrer's formula and was found in the range of 36-43 nm which confirms the nano crystalline form. The lattice constant of the samples increases with cobalt composition, obeys the Vegard's law [16].

Table 1. Structural parameters of the prepared Li-Co ferrite samples.

Composition	Mol.wt (gm/mol)	Crystallite Size D_{xrd} (nm)	Lattice constant (Å)	X-ray Density (d_x) (gm/cc)	Experimental Density (d_e) (gm/cc)	% of Porosity (P)
$Li_{0.5}Fe_{2.5}O_4$	207.079	41.90	8.35	4.71	4.28	9.00
$Li_{0.4}Co_{0.2}Fe_{2.4}O_4$	212.587	43.01	8.37	4.81	4.37	9.10
$Li_{0.3}Co_{0.4}Fe_{2.3}O_4$	218.095	38.44	8.37	4.93	4.38	11.00
$Li_{0.2}Co_{0.6}Fe_{2.2}O_4$	223.603	37.57	8.38	5.03	4.39	12.70
$Li_{0.1}Co_{0.8}Fe_{2.1}O_4$	229.111	37.06	8.39	5.14	4.58	10.90
$CoFe_2O_4$	234.619	36.90	8.40	5.25	4.68	10.70

From Fig. 2(a) one can observe that the lattice constant of the prepared samples increases with the increase in cobalt content (x), probably due to the ionic radius of the six-fold coordinated Co^{+2} (0.745 Å) being larger than the average radius of the six-fold coordinated Fe^{3+} (0.645 Å) and Li^{+1}(0.76 Å) [17]. X-ray density of the prepared samples increases with cobalt concentration as shown in Fig. 2(b), because of the larger molar mass of the substituted cobalt ions compared with iron and lithium ions. From table 1 one can conclude that the experimental densities of the samples were observed to be less than the X-ray density which reveals the porous nature of the prepared samples [18].

The distance between magnetic ions (hopping length) in 'A' (Tetrahedral-d_A) and 'B' sites (Octahedral-d_B) were calculated using relations (eqs.1) and the values are given in table 2.

$$d_A = 0.25a\sqrt{3}, \text{ and } d_B = 0.25a\sqrt{2}$$

(1)

The relation between hopping length for tetrahedral and octahedral sites as a function of Co content (x) is shown in Fig. 3. The distance between the magnetic ions in the lattice depends on the lattice constant of the prepared sample.

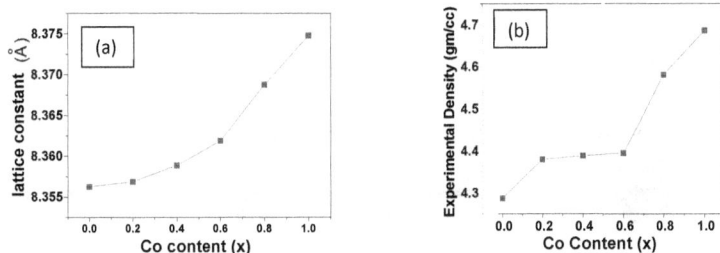

Fig. 2 (a). Variation of Lattice constant with Co content (x).

Fig. 2 (b). Variation of experimental density with Co content (x).

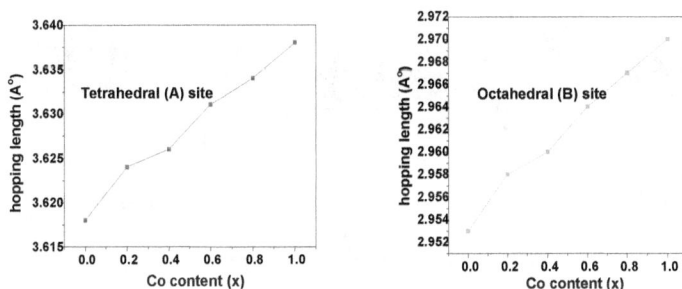

Fig. 3. Variation of hopping length in tetrahedral (A) and octahedral (B) sites with Co content.

Table 2. Values of hopping length for A-site (d_A) and B-site (d_B) of Li-Co ferrites

Sr.No	Composition	(d_A) in Å	(d_B) in Å	Unit cell Volume ($Å^3$)
1	$Li_{0.5}Fe_{2.5}O_4$	3.618	2.953	583.48
2	$Li_{0.4}Co_{0.2}Fe_{2.4}O_4$	3.624	2.958	586.39
3	$Li_{0.3}Co_{0.4}Fe_{2.3}O_4$	3.626	2.960	587.36
4	$Li_{0.2}Co_{0.6}Fe_{2.2}O_4$	3.631	2.964	589.82
5	$Li_{0.1}Co_{0.8}Fe_{2.1}O_4$	3.634	2.967	591.58
6	$CoFe_2O_4$	3.638	2.970	593.55

3.2 SEM micrographs

The scanning electron microscopic (SEM) images of the cobalt substituted lithium ferrite nano-particles, sintered at 500 °C for 4 h are shown in Fig. 4. The SEM micrograph shows the surface morphology and grain size of the prepared samples. It is clear from the micrographs that prepared samples are uniformly distributed and less agglomerated.

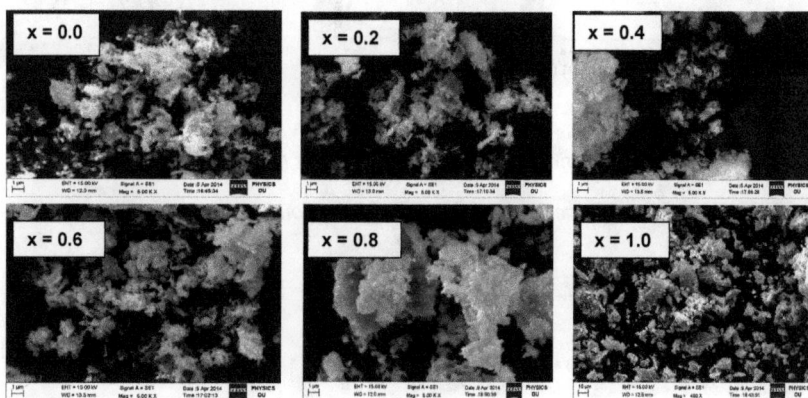

Fig. 4. SEM micrographs of nano crystalline $[Li_{0.5}Fe_{0.5}]_{1-x}Co_x Fe_2O_4$ (x = 0.0 to 1.0) ferrites

The homogeneity in the shape and the grain size largely affects the electrical and magnetic properties of the investigated ferrites. In Li-Co ferrites, cobalt substitution brings uniform microstructure throughout composition and a smaller increase in grain size is observed with the increase in Co substitution.

3.3 Energy dispersive spectroscopic analysis

Energy dispersive x-ray spectroscopy (EDS, EDX or XEDS) is an analytical method used for the elemental analysis or chemical characterization of a sample. It depends on the investigation of an interaction of some source of x-ray excitation and prepared sample. Figure 5 show the EDS pattern obtained for prepared samples of $[Li_{0.5}Fe_{0.5}]_{1-x}Co_x Fe_2O_4$ ferrites which gives the elemental and atomic compositions in the prepared samples. The element Li cannot be detected due to its low atomic weight by this EDS measurement due to the limitation of the used equipment [19].

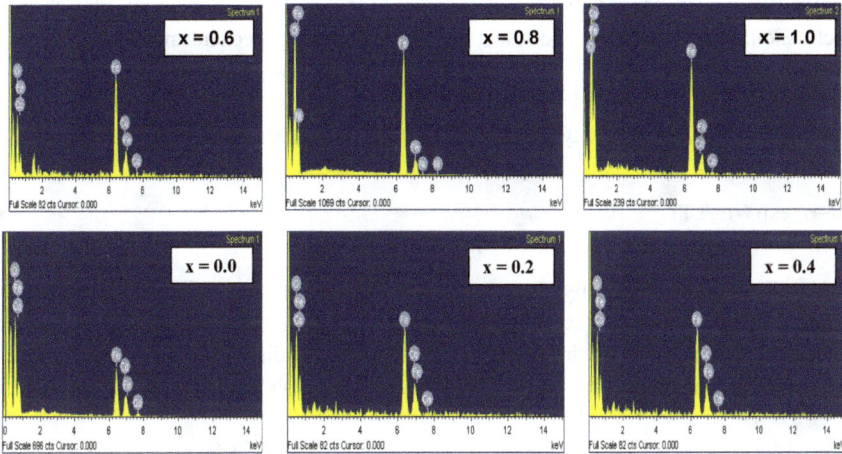

Fig. 5. EDS spectra of nano crystalline $[Li_{0.5}Fe_{0.5}]_{1-x}Co_x Fe_2O_4$ (x = 0.0 to 1.0) ferrites.

3.4   DC electrical resistivity studies of $[Li_{0.5}Fe_{0.5}]_{1-x}Co_x Fe_2O_4$ nano ferrites

DC electrical resistivity studies of $[Li_{0.5}Fe_{0.5}]_{1-x}Co_x Fe_2O_4$ nano crystalline ferrites with different composition (x) were measured in the temperature ranges from 200 °C to 600 °C using the two probe method (temperature increment step of 10 °C). An electrical resistance (R) values of various compositions (x) were obtained in temperature ranges from 200 to 600 °C. The DC electrical resistivity (ρ) was calculated using eqs. 2.

$$\rho = \frac{RA}{l} \qquad (2)$$

Where R is the resistance, l is thickness and A is the cross section area of the sample.

The DC electrical resistivity is the important property to study the conduction mechanism in ferrites. The conduction mechanism in ferrites is due to the hopping of charge carriers (electrons), between the Fe^{2+} and Fe^{3+} ions of same element, distributed randomly over equivalent crystallographic sites in the lattice. The probability of hopping among charge carriers depends on the separation between the involved ions and activation energy [20]. The electrostatic interaction between the conduction electron and nearby ions may result in the polarization of the surrounding region so the electron present at the center of the polarization well. This electron is transferred to the neighboring site by thermal activation. This mechanism in conduction is called hopping mechanism [21].

3.4.1 Effect of temperature on dc conductivity

The dc electrical conductivity [22, 23] of the of ferrite materials has the general form

$$\sigma = \sigma_o \exp(-E_a/kT) \tag{3}$$

where E_a is the thermal activation energy, σ_o is the pre exponential factor depending on material nature and k is the Boltzmann constant. The temperature dependence of the electrical conductivity of the prepared samples was studied from Arrhenius plots by plotting a graph between the log (σT) vs 1000/T which is shown in Fig. 6.

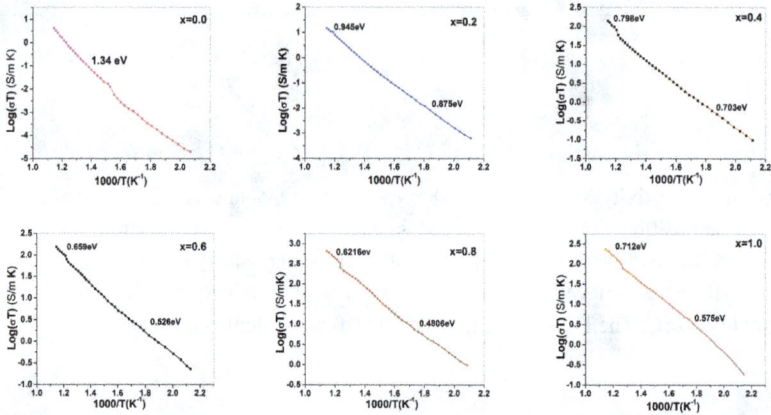

Fig. 6. Arrhenius plots for electrical conductivities of nano crystalline Li-Co ferrites.

It yields a straight line whose slope can be used to calculate thermal activation energies of the ferrite samples. From Fig. 6, it is clear that the conductivity of the ferrite samples increases with increase in the temperature. i.e. as the temperature increases, resistivity of the prepared ferrite samples decreases, indicating the semiconducting behavior. All the plots (except pure lithium ferrites) of electrical conductivity log (σT) vs 1000/T yield a change in slope at a particular temperature. This change in slope occurs while crossing the Curie temperature (the temperature at which the ferrimagnetic material changes to paramagnetic). The discontinuity at the Curie temperature was attributed to the magnetic transition from well ordered ferrimagnetic state to disordered paramagnetic state which involves different activation energies. The values of the electrical resistivity and thermal activation energies of the prepared samples in ferrimagnetic region and paramagnetic region were given in table 3. From the slope of the Arrhenius plots log ρ vs 1000/T of various compositions of the Li-Co nano ferrites shown in Fig. 6, activation energy (Ea) was calculated using eqs. 4.

$$E_a = 2.303 \times k_B \times 10^3 \times slope \ (eV) \qquad (4)$$

Table 3. Curie temperature and Activation energies of Li-Co ferrites.

Sr.No.	Composition	Curie Temp ($^\circ$C)	E_a in Paramagnetic region (eV/mol)	E_a in Ferromagnetic region (eV/mol)
1	$Li_{0.5}Fe_{2.5}O_4$	---	---	1.34
2	$Li_{0.4}Co_{0.2}Fe_{2.4}O_4$	572	0.945	0.875
3	$Li_{0.3}Co_{0.4}Fe_{2.3}O_4$	560	0.798	0.703
4	$Li_{0.2}Co_{0.6}Fe_{2.2}O_4$	553	0.659	0.526
5	$Li_{0.1}Co_{0.8}Fe_{2.1}O_4$	540	0.621	0.480
6	$CoFe_2O_4$	521	0.712	0.575

From Fig. 6, one can conclude that except of pure lithium ferrite, Arrhenius plots for electrical conductivities of cobalt substituted lithium ferrites shows a discontinuity (kink occurs) in a straight line, which shows two different regions with a variation in the activation energies. The temperature where the discontinuous kink occurs is called transition temperature (T^*). V.R.K. Murthy et al. [24] reported that discontinuity in the Arrhenius plot may be due to the change in the conduction mechanism. For $T < T^*$ region, the conduction mechanism may be due to the electron hopping between the Fe^{2+} and Fe^{3+} ions in the octahedral sites of the crystal lattice. For $T > T^*$ region the conduction mechanism may be probably due to the ionic conduction caused by the lithium and cobalt ions in the B-sites. There exist valence state transitions in Fe and Co ions as mentioned below [25].

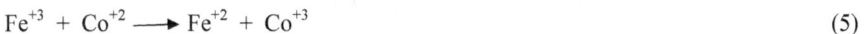

$$Fe^{+3} + Co^{+2} \longrightarrow Fe^{+2} + Co^{+3} \qquad (5)$$

According to Raghasudha et al. [26], discontinuity in the Arrhenius plot may be due to the ferri to para magnetic transition from low to high temperature. This transition temperature (T^*) represents the Curie temperature (T_c). Therefore it is suggested that the change in slope is due to the magnetic transition. Similar transitions near the Curie temperature have also been observed by several investigators for different ferrites [27, 28]. From table 3 it is revealed that activation energy of the paramagnetic region is larger than the ferrimagnetic region. This is because ferrimagnetic state was an ordered state while the paramagnetic state is a disordered state. Hence, charge carriers require more energy for conduction in paramagnetic state compared to ferrimagnetic state.

3.4.2 Effect of cobalt substitution on dc resistivity of $[Li_{0.5}Fe_{0.5}]_{1-x}Co_x Fe_2O_4$

The variation of DC resistivity with cobalt concentration (x) is shown in Fig. 7. It is clear from Fig. 7 that DC resistivity of the prepared Li-Co samples decreases with Co composition. This is because the hopping of electrons between the $Fe^{2+} \leftrightarrow Fe^{3+}$ and $Co^{2+} \leftrightarrow Co^{3+}$ increases in the octahedral sites of the prepared samples with increase in the Co composition [29]. From this one can reveal that the pure ferrites have more resistivity than the substituted ferrites i.e. by substituting the resistivity of ferrites decreases. The decrease in resistivity of Li-Co ferrites with Co composition can also be due to the increase in porosity of the samples. The variations of resistivity with cobalt composition at different temperatures are given in the table 4.

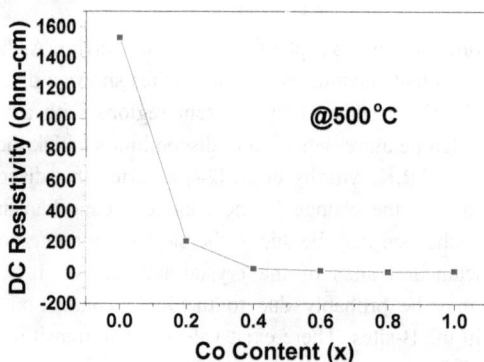

Fig. 7. Variation of DC resistivity of $[Li_{0.5}Fe_{0.5}]_{1-x}Co_x Fe_2O_4$ with Co content (x).

Table 4. Resistivity at different temperatures of nano crystalline Li-Co ferrites

Composition	DC resistivity (ohm-cm)			
	600 °C	500 °C	400 °C	200 °C
$Li_{0.5}Fe_{2.5}O_4$	152.47	1528	20235	9.21×10^8
$Li_{0.4}Co_{0.2}Fe_{2.4}O_4$	41.48	205	1501	329490
$Li_{0.3}Co_{0.4}Fe_{2.3}O_4$	6.156	27	99	17870
$Li_{0.2}Co_{0.6}Fe_{2.2}O_4$	5.640	14.29	48.96	11389
$Li_{0.1}Co_{0.8}Fe_{2.1}O_4$	1.410	7.35	25.81	1643
$CoFe_2O_4$	4.320	10.45	32.31	16783

It is clear from table 4 that by decreasing the temperature, the resistivity of the prepared samples increases and by increasing the cobalt composition in the Li-Co ferrite, the resistivity decreases. A similar behavior is observed in other ferrites [24].

3.5   Dielectric studies of $[Li_{0.5}Fe_{0.5}]_{1-x} Co_x Fe_2O_4$ nano ferrites

3.5.1   Effect of temperature on dielectric constant of Li-Co ferrite

The temperature dependence of the dielectric constant of the prepared $[Li_{0.5}Fe_{0.5}]_{1-x} Co_x Fe_2O_4$ ferrites is shown in figure 8. From the figure, one can conclude that except for pure lithium ferrite, dielectric constant of the cobalt substituted lithium ferrite increases slowly in the beginning (300 to 450 K), stabilizes up to a certain temperature (475 K), and then increases rapidly thereafter. In case of pure lithium ferrites the dielectric constant was increased up to certain temperature and beyond that temperature a decrease in dielectric constant was observed [29].

Similar behavior is observed by other researchers for different compositions [30]. The hopping of electrons between the Fe^{2+} and Fe^{3+} ions present in the octahedral sites is thermally activated by increasing temperature which causes local displacements in the direction of the external field, which enhances their contribution to the space charge polarization thereby leading to increase in the dielectric constant [31]. In general, the dielectric constant of any material arises due to the dipolar, ionic, electronic and interfacial polarizations. At low frequencies, dipolar and interfacial polarizations are known to play dominant role and these are dependent on temperature. The rapid increase in the dielectric constant with temperature at low frequencies comes from the combined effect of interfacial and dipolar polarizations. At high frequencies the ionic and electronic polarizations are main contributors and these are independent of temperature [32]. The present study reports that dielectric characteristics of Li-Co samples prepared by an auto combustion method have low dielectric constant compared to earlier reported Li-Co samples, prepared by ceramic method, had high dielectric constants [33].

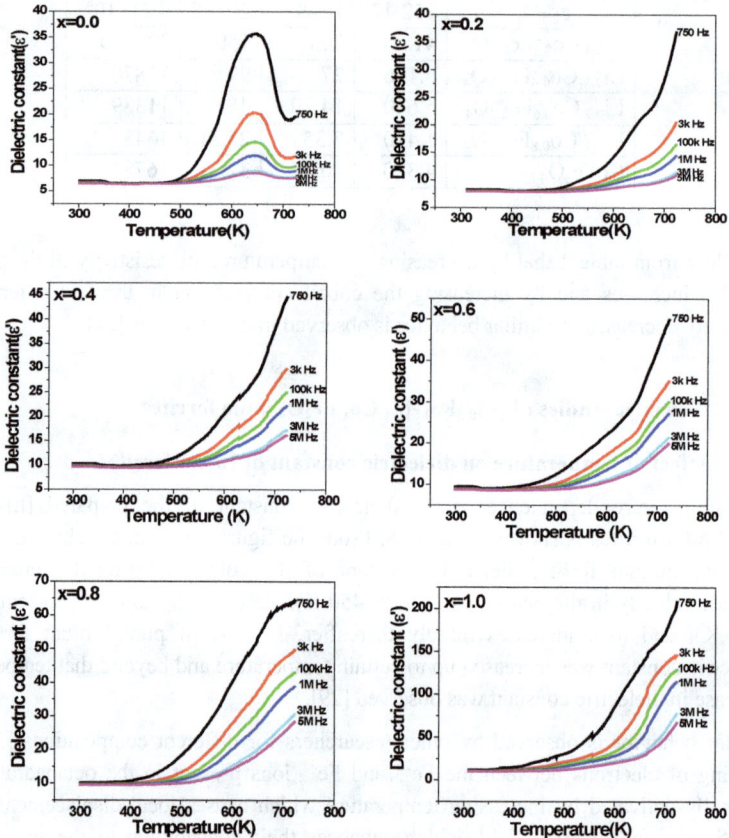

Fig.8. Variation of (ὲ) with temperature at different frequencies of $[Li_{0.5}Fe_{0.5}]_{1-x}Co_xFe_2O_4$.

3.5.2 Effect of temperature on dielectric loss tangent of Li-Co ferrites

The variation of dielectric loss tangent (tan δ) with temperature of the prepared samples at different definite frequencies are shown in Fig.9. This behavior is similar to that of variation of dielectric constant with temperature as shown in Fig. 8. The dielectric loss tangent measures the energy loss within the prepared material. This loss arises when the polarization lags behind the applied AC electric field and this is caused due to the impurities and imperfection in the crystal lattice. The increase in dielectric loss tangent with temperature is due to the increased conduction of thermally activated electrons [34].

From Fig. 8 and Fig. 9 one can conclude that by increasing the temperature, both the dielectric constant and dielectric loss tangent of the prepared nano crystalline Li-Co ferrite samples increased.

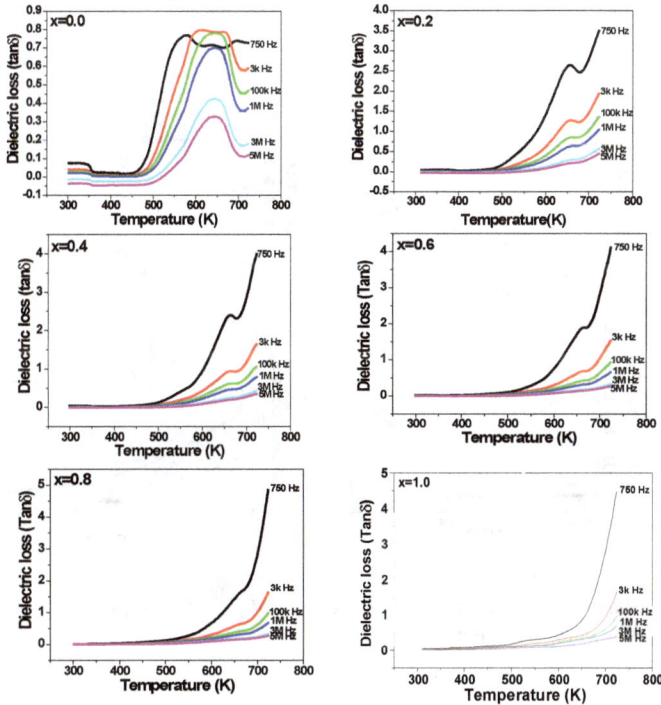

Fig. 9. Variation of Dielectric loss of Li-Co nano ferrites with temperature at different frequencies.

3.5.3 Frequency dependence of dielectric constant of Li-Co nano ferrites at different temperature

Dielectric properties of the ferrite materials depend on the temperature, frequency of the applied electric field and its structure [35]. Fig. 10 shows the observed frequency dependence of dielectric constant of prepared samples.

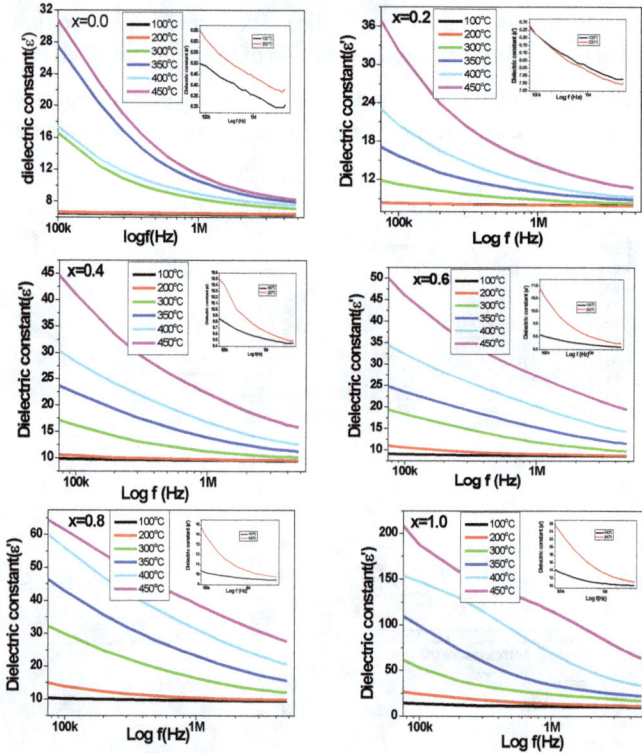

Fig.10. Variation of dielectric constant (real-ε') of $[Li_{0.5}Fe_{0.5}]_{1-x} Co_x Fe_2O_4$ nano ferrites with frequency at different temperature.

From Fig. 10, it is observed that dielectric constant decrease rapidly in the low frequency region whereas the decrement is very slow in the high frequency region. This type of nature can be explained by the Maxwell-Wagner and Koop's phenomenological theory

[36]. According to this theory, the dielectric medium is assumed to be made up of highly conducting grains which are separated by non conducting grain boundaries. The non conductive grain boundaries are more effective at low frequencies whereas the conductive grains are observed to be more effective at high frequencies. In ferrites, motion of electrons takes place between the Fe^{2+} and Fe^{3+} ions present at crystallographically equivalent sites i.e. B-sites. This phenomenon is called hopping mechanism of electron transfer. The decrease in dielectric constant with increase in frequency is due to the fact that the electron hopping between the Fe^{2+} and Fe^{3+} ions at the B-sites cannot follow the change in the AC electric field at high frequencies. The electrons have to pass through the good conducting grains and non conducting grain boundaries. As the grain boundaries have large resistance so the electrons pile up there and produce large space charge polarization.

Therefore, the dielectric constant has large value at lower frequency range. With further increase in frequency the electrons change their direction of motion, which hinders the motion of electrons inside the material so the accumulation of electrons at the grain boundaries decreases. This decreases the space charge polarization. Therefore the dielectric constant also decreases. Low temperature (100 and 200 °C) variations were indicated separately in the inset of Fig. 10.

3.5.4 Dependence of dielectric constant of Li-Co nano ferrites on cobalt composition

From Fig.11, it was found that dielectric constant (ε') for LC00 > LC02 (x = 0.0 > x = 0.2) of the prepared samples.

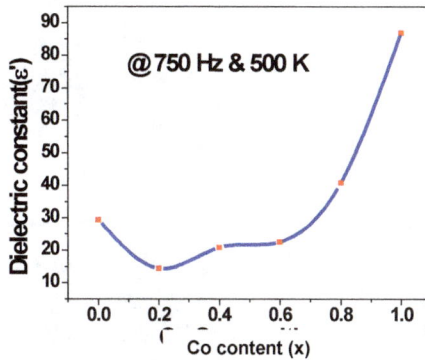

Fig. 11. Effect of Co content (x) on Dielectric constant of $[Li_{0.5}Fe_{0.5}]_{1-x}Co_x Fe_2O_4$ ferrites.

Dielectric constant (ϵ') is observed to decrease with Co content up to x = 0.2 and then further it is found to increase with increasing cobalt content. The lowest value of dielectric constant (ϵ') for composition is observed at x = 0.2. Because, cobalt substitution decreases the number of Fe^{2+} ions in the B-site and hinders the interaction between Fe^{2+} and Fe^{3+} hence dielectric constant decreases. Further substitution of cobalt ions (x > 0.2) probably gives rise to hopping of electrons between Co^{2+} and Co^{3+} ions which increases the dielectric constant [37].

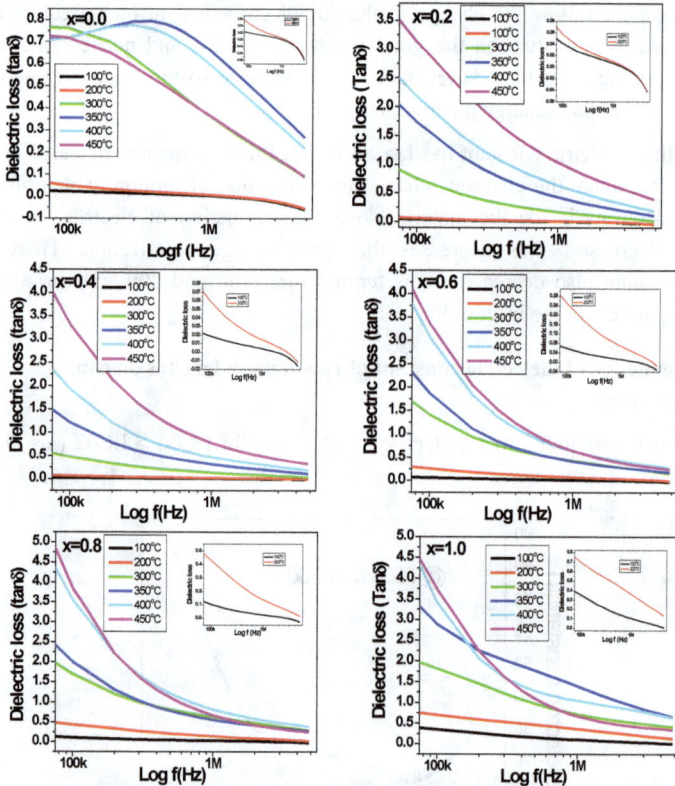

Fig. 12. Variation of dielectric loss tangent of $[Li_{0.5}Fe_{0.5}]_{1-x}Co_xFe_2O_4$ nano ferrites with frequency at different temperature

3.5.5 Frequency dependence of dielectric loss (tan δ) of Li-Co nano ferrites at different temperature

Fig. 12 shows the frequency dependence of dielectric loss tangent at different temperatures for prepared ferrite samples.

From these figures, it is observed that the dielectric loss tangent decreases as the frequency of the applied electric field increases because the hopping frequency of electrons cannot follow the frequency of the applied electric field after certain frequency. Low temperature (100 and 200° C) variations are indicated separately in the inset of Fig. 12 for clear observation.

3.6 Magnetic properties of $[Li_{0.5}Fe_{0.5}]_{1-x}Co_x\,Fe_2O_4$ ferrites

3.6.1 Hysteresis loops of Li-Co ferrites

The magnetic measurements of various compositions of heat treated Li-Co nano ferrite samples were measured by using vibrating sample magnetometer (VSM, Model no.155) at room temperature under an applied field of 20 kOe.

Fig. 13 shows the dependence of magnetization values (M) on the applied magnetic field (H), (Hysteresis loops) of $[Li_{0.5}Fe_{0.5}]_{1-x}Co_x\,Fe_2O_4$ ferrites .The magnetization values in the presence of applied magnetic field for fabricated samples exhibit a clear hysteresis loop behavior and various magnetic parameter values are given in the table 5. The shape and width of the hysteresis loop is influenced by several factors including chemical composition, method of fabrication, sintering temperature/time and grain size etc. In spinel ferrites the magnetic moment of the A-site and B-site were aligned anti-parallel and shows a ferri-magnetism with a magnetization $M\,(M_B-M_A)$.

Table 5. Magnetic parameters (M_s, H_c and M_r) values of the prepared nano crystalline Li-Co Samples.

Composition	M_s (emu/g)	Coercivity H_c (Oe)	M_r (emu/g)	M_r/M_s	Magnetic moment (μ_B)	Anisotropy constant (K) (erg/cm^3)
$Li_{0.5}Fe_{2.5}O_4$	55.93	153	15.15	0.27	2.0	8731
$Li_{0.4}Co_{0.2}Fe_{2.4}O_4$	58.30	888	26.18	0.45	2.2	26253
$Li_{0.3}Co_{0.4}Fe_{2.3}O_4$	60.68	424	30.08	0.50	2.3	52826
$Li_{0.2}Co_{0.6}Fe_{2.2}O_4$	62.39	1388	30.31	0.49	2.4	88364
$Li_{0.1}Co_{0.8}Fe_{2.1}O_4$	68.29	1543	34.36	0.50	2.8	107521
$CoFe_2O_4$	75.40	2099	31.15	0.41	3.2	121742

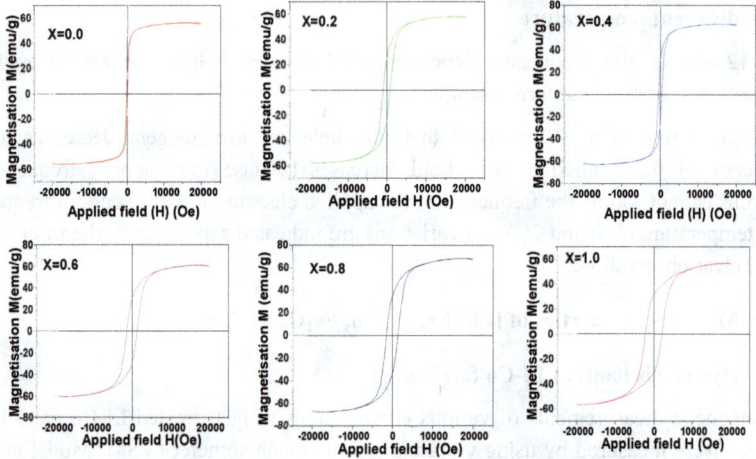

Fig.13. Hysteresis loops of nano crystalline $[Li_{0.5}Fe_{0.5}]_{1-x}Co_xFe_2O_4$ (x = 0.0 to 1.0) ferrites.

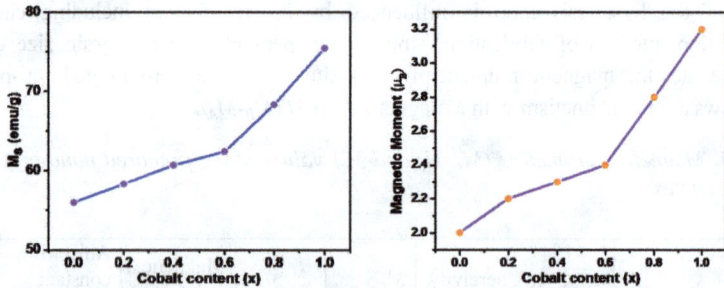

Fig.14. Variation of Saturation magnetization (a) and Magnetic moment (b) with Co content (x) for $[Li_{0.5}Fe_{0.5}]_{1-x}Co_xFe_2O_4$ (x = 0.0 to 1.0) ferrites.

It is clear from Table 5 and Fig 14 (a), saturation magnetization (M_s) is found to increase with increase with Co substitution in the Li-Co ferrites. This can be explained on the basis of cation distribution. It is reported that Li^+ ions occupy only B-site whereas Co^{2+}

ions can occupy both A and B sites [38] and Co^{2+} ions have more magnetic moment ($3\mu_B$) [35] than the Li^+ (0 μ_B i.e. non magnetic ion) and less magnetic moment than Fe^{3+} ion ($5\mu_B$) [39]. When cobalt with magnetic moment $3\mu_B$ is substituted in lithium ferrites the resulting magnetic moment is increased as evident from table 5 and shown in Fig. 14. From the Fig. 13, one can observe that pure lithium ferrite was soft ferrite and pure cobalt ferrite was hard ferrite. So, by the doping of cobalt in the lithium ferrite, the system changes from the soft ferrites in to the hard ferrites.

Coercivity is the magnetic field strength required for overcoming anisotropy to flip the magnetic moment which is influenced by the doping [35]. The coercivity values were in the range of 153-2099 Oe. The magnetic coercivity of the materials depends on the magneto-crystalline anisotropic energy, micro-strain, inter-particle interaction, temperature size etc [40]. For cobalt ferrites the Co^{2+} ($3d^7$, $4F_{9/2}$, L=3, S=3/2, J=9/2) cations possess 7 d electrons, three of which are unpaired. Evidently large magneto crystalline anisotropic energy may be due to the strong L-S coupling on the Co^{2+} cation sites. Energy barriers in the individual constituents was considered and introduced by the magneto crystalline anisotropic energy of the prepared cobalt ferrite [41]. From the table 5, one can observe that magnetic moment increased with increase in the cobalt concentration in the Li-Co ferrites.

From all these results, it is clear that by increasing the cobalt concentration in the nano crystalline Li-Co ferrite, magnetization of the samples increases and the material is being converted from soft magnetic to finally hard magnetic. Accordingly, among all compositions the one with x = 1.0 have high coercivity hence, can be used for fabrication of hard permanent magnets. Further, based on the coercivity and remanence values of doped ferrites, they could be more suitable for magnetic recording purposes. The magnetization observed for prepared Li-Co ferrites by auto combustion method was higher than the values obtained by the standard ceramic method [42].

3.6.2 (FC) and (ZFC) magnetization study of $CoFe_2O_4$ and $Li_{0.1}Co_{0.8}Fe_{2.1}O_4$ nano ferrites

Temperature dependent magnetic properties of $[Li_{0.5}Fe_{0.5}]_{1-x}Co_xFe_2O_4$ for two compositions with cobalt content x = 0.8 and x = 1.0 were carried out using Vibrating sample magnetometer (VSM). The magnetization as a function of an applied field ±10T was carried out at temperatures 5 K and 310 K. Field cooled (FC) and Zero field cooled (ZFC) magnetization measurements under an applied field of 100 Oe and 1 kOe in the temperature range of 5K to 375K were performed. These measurements result the blocking temperature (T_b) at around 350K i.e. above room temperature for both the ferrites. Below this temperature the ferrites show ferrimagnetic behavior and above which superparamagnetic behavior where the coercivity and remanence magnetization are almost zero. Such behavior makes the ferrites to be desirable for biomedical applications.

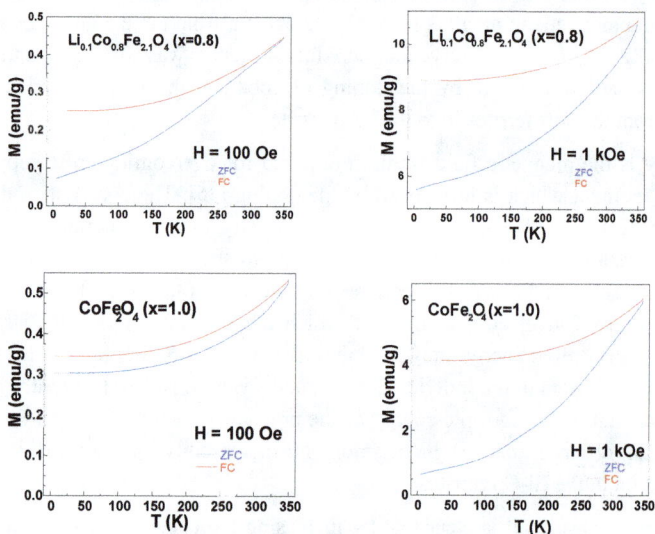

Fig. 15. Magnetization-Temperature curves recorded in FC and ZFC modes for the sample $Li_{0.1}Co_{0.8}Fe_{2.1}O_4$ and $CoFe_2O_4$ in an external magnetic field of 100 Oe and 1 kOe.

Among the prepared lithium-cobalt nano ferrites of all compositions, the samples with cobalt content x = 0.8 and x = 1.0 have the crystallite size of 37.06 nm and 36.9 nm. Ferrites with such low particle size are expected to show superparamagnetic behavior. This has motivated the author to investigate the superparamagnetic behavior of these samples by performing Zero Field Cooled (ZFC) and Field Cooled (FC) magnetization measurements using Vibrating Sample Magnetometer.

Superparamagnetism is a phenomenon in which the magnetic materials behave as paramagnetic materials below the blocking temperature unlike the general transition of a magnetic material from ferromagnetic to paramagnetic above its Curie temperature. In the ZFC process, the ferrite sample is cooled in the absence of magnetic field. Then, the temperature is gradually raised by applying a moderate measuring field and the magnetization values (*M*) were recorded. In the FC process, the sample is cooled in the presence of a nonzero magnetic field and the same procedure is followed as in case of ZFC process. Fig. 15 shows the Magnetization-Temperature curves recorded in FC and ZFC modes for the samples $Li_{0.1}Co_{0.8}Fe_{2.1}O_4$ and $CoFe_2O_4$ in an external magnetic field of 100 Oe and 1 kOe respectively.

In ZFC mode, the two ferrite samples under investigation i.e. $Li_{0.1}Co_{0.8}Fe_{2.1}O_4$ and $CoFe_2O_4$ were cooled from 350 K down to 2 K in the absence of magnetic field. Then, a measuring field of 100 Oe and 1 kOe were applied and the magnetization measurements were made in heating cycle. Whereas, in FC mode the samples were cooled from 350 K down to 2 K in the presence of the measuring field and then magnetization measurements were recorded as function of rising temperature. From the figures it is clear that both the FC and ZFC magnetization decrease by decreasing the temperature for both the samples under two different applied fields (100 Oe and 1 kOe).

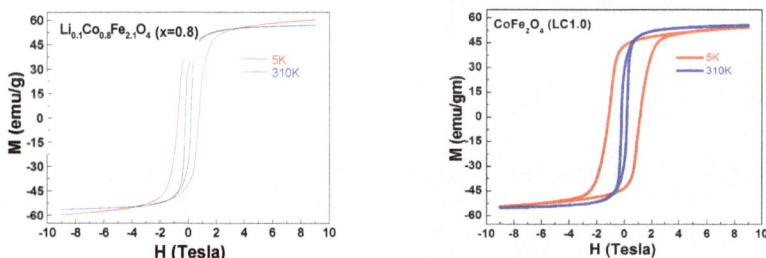

Fig. 16. Hysteresis loops for $Li_{0.1}Co_{0.8}Fe_{2.1}O_4$ and $CoFe_2O_4$ at 5 K and 310 K

There exists a bifurcation between the FC and ZFC modes as evident from the figures owing to the magnetic relaxation nature of the nano particles and confirms their (SPM) super-paramagnetic nature. The temperature at which this bifurcation in the two modes is observed is defined as bifurcation temperature or blocking temperature (T_b). It is observed that the blocking temperature did not change with the increase in applied field but shows a strong bifurcation in the FC and ZFC curves under higher applied field i.e. at 1 kOe, showing irreversibility of FC and ZFC magnetization curves. For both the samples the bifurcation or blocking temperature is observed at around 350 K.

Fig. 16 shows Magnetization Hysteresis loops for $Li_{0.1}Co_{0.8}Fe_{2.1}O_4$ and $CoFe_2O_4$ samples at 5 K and 310 K. It is observed that the coercivity is more at lower temperature and it decreases with increase in temperature. The values of Coercivity (H_c) and remanence (M_r) were measured from the hysteresis curves and were recorded in table 6.

Table 6. Coercivity (H_c) and Remanance (M_r) of $Li_{0.1}Co_{0.8}Fe_{2.1}O_4$ and $CoFe_2O_4$ at 5K and 310K.

Magnetic parameters	$Li_{0.1}Co_{0.8}Fe_{2.1}O_4$		$CoFe_2O_4$	
	5K	310K	5K	310K
(H_c in Oe)	0.71T	0.18T	1.09T	0.19T
(M_r in emu/g)	39.46	30.62	43.30	30.99

From this table 6, it is clear that H_c and M_r values at 310 K for both the samples is very less which will approach zero at above room temperature. Zero coercivity and zero remanence is the characteristic feature of superparamagnetic (SPM) behavior of the magnetic nano particles [43]. By comparing the FC-ZFC data and hysteresis curves data it is clear that below blocking temperature i.e. < 350K the material shows some hysteresis and hence behaves as ferromagnetic material. Above blocking temperature i.e. > 350 K, hysteresis disappears and the material behaves as superparamagnetic. Hence, Li-Co nano ferrites with superparamagnetic behavior are desirable for bio-medical applications.

Conclusions

Cobalt substituted Lithium nano ferrites with the chemical composition $Li_{0.5-0.5x}Co_xFe_{2.5-0.5x}O_4$ (x = 0.0 to 1.0, with a step increment of 0.2) were prepared by citrate gel auto combustion method. X-ray diffraction analysis confirms the formation of single phased cubic spinel structure of the ferrites with an average crystallite size in the range of 36-42 nm. Scanning Electron Microscopic (SEM) analysis indicates that the grains are uniformly distributed and less agglomerated. Energy Dispersive Spectroscopic (EDS) analysis shows the presence of Co, Fe, O and Lithium element was not detected due to low atomic weight. DC resistivity of all the samples suggests the semiconducting behavior. Dielectric constant and dielectric loss tangent of all samples decreased as a function of frequency. Pure lithium ferrite is a soft ferrite whereas cobalt doped lithium ferrite resulted into a hard ferrite. Li-Co ferrites with x = 0.8 and x = 1.0 compositions show superparamagnetic behavior. This property can be used in biomedical applications for targeted drug delivery and magnetic resonance imaging (MRI).

References

[1]   S.C. Watawe, U.A. Bamme, S.P. Gonbare, R.B. Tangsali, Preparation and dielectric properties of Li-Cd ferrite using microwave-induced combustion, Material Chemistry Physics, 103 (2007) 323-328. https://doi.org/10.1016/j.matchemphys.2007.02.037

[2]   S.T. Assar, H.F. Abosheiasha, M.K. El Nimr, Comparison study of the magnetic permeability and dc conductivity of Co-Ni-Li ferrite nano particles and their bulk counterparts, Journal of Magnetism and Magnetic Materials, 354 (2014) 1-6. https://doi.org/10.1016/j.jmmm.2013.10.029

[3]   P. Venugopal Reddy, Charge transport in Li-Ni ferro spinels, Journal of Applied Physics, 63(1988) 3783-3785. https://doi.org/10.1063/1.340639

[4]   H.P.J. Wijn, E.W. Gorter, C.J. Esveldt, P. Geldermans, Conditions for square hysteresis loops in ferrites, Philips Technical Review,16 (1954) 49-58.

[5]   Rajesh Cheruku, G. Govindaraj, Lakshmi Vijayan, Super-linear frequency dependence of ac conductivity in nano crystalline lithium ferrite, Materials Chemistry Physics,146 (2014) 389-398. https://doi.org/10.1016/j.matchemphys.2014.03.043

[6]   S.T. Assar, H.F. Abosheiasha, M.K. El Nimr, Study of the dielectric behavior of Co-Ni-Li nanoferrites, Journal of Magnetism and Magnetic Materials, 350 (2014) 12-18. https://doi.org/10.1016/j.jmmm.2013.09.022

[7]   M. Raghasudha, D. Ravinder, P. Veerasomaiah, TIR Studies and Dielectric Properties of Cr Substituted Cobalt Nano Ferrites Synthesized by Citrate-Gel method, Nanoscience and Nanotechnology, 3(2013)105-111.

[8]   Mathew George, Swapna S. Nair, Asha Mary John, P.A. Joy, M.R. Anantharaman, Structural, magnetic and electrical properties of sol-gel prepared $Li_{0.5}Fe_{2.5}O_4$ fine particles, Journal of Physics D: Applied Physics, 39 (2006) 900-910. https://doi.org/10.1088/0022-3727/39/5/002

[9]   K.S. Aneesh Kumar, R.N. Bhowmik, Microstructural characterization and magnetic study of nickel ferrite synthesized through coprecipitation route at different pH values, Materials Chemistry and Physics,146 (2014) 159-169. https://doi.org/10.1016/j.matchemphys.2014.03.015

[10]   Sangeeta Thakur, S.C. Katyal, M. Singh, Structural and magnetic properties of nano nickel–zinc ferrite synthesized by reverse micelle technique, Journal of Magnetism and Magnetic Materials, 321(2009)1-7. https://doi.org/10.1016/j.jmmm.2008.07.009

[11]   C. Liu, B. Zou, AJ. Rondinone, ZJ. Zhang, Chemical control of super paramagnetic properties of magnesium and cobalt spinel ferrite nano particles through atomic level magnetic coupling, Journal of the American Chemical Society,122 (2000) 6263-6267. https://doi.org/10.1021/ja000784g

[12] S. Modak, S. Karan, S.K. Roy, S. Mukherjee, D. Das, P.K. Chakrabarti, Preparation and characterization of SiO_2 coated nano particles of Mn-Zn ferrites, Journal of Magnetism and Magnetic Materials, 321 (2009) 169-174. https://doi.org/10.1016/j.jmmm.2008.08.073

[13] M.S. Ruiz, S.E. Jacobo, Electromagnetic properties of Li-Zn ferrites doped with aluminum, Physica B, 407 (2012) 3274-3277. https://doi.org/10.1016/j.physb.2011.12.085

[14] Mamatha Maisnam, Sumitra Phanjoubam, Frequency dependence of electrical and magnetic properties of Li-Ni-Mn-Co ferrites, Solid State Communication, 152 (2012) 320-323. https://doi.org/10.1016/j.ssc.2011.11.019

[15] S.A. Saafan, S.T. Assar, B.M .Moharram, M.K. El Nimr, Comparison study of some structural and magnetic properties of nano structured and bulk Li-Ni-Zn ferrite samples, Journal of Magnetism and Magnetic Materials, 322 (2010) 628-632. https://doi.org/10.1016/j.jmmm.2009.10.027

[16] K. Wykpis, A. Budnoik, E. lagiewka, The Influence of Sodium Molybdate on the Properties of Zn-Ni Layers Obtained by Electrolytic Deposition, Material Science Forum, 636 (2010)1053-1058. https://doi.org/10.4028/www.scientific.net/MSF.636-637.1053

[17] R.D. Shannon, Revised effective ionic radii and systematic studies of interatomic distances in halides and chalcogenides, Acta Crystallographica section A, 32 (1976) 751-767. https://doi.org/10.1107/S0567739476001551

[18] K. Rama Krishna, D. Ravinder, K. Vijay Kumar, Utpal S Joshi, V.A. Rana, Ch. Abraham lincon, Dielectric Properties of Ni-Zn Ferrites Synthesized by Citrate Gel method, World Journal Condensed Matter Physics, 2 (2012) 57-60. https://doi.org/10.4236/wjcmp.2012.22010

[19] Keqiang Ding, Jing Zhao, Mian Zhao, Yuying Chen, Yongbo Zhao, Jinming Zhou, The Effect of Ti Doping on the Electrochemical Performance of Lithium Ferrite, International Journal of Electrochemical Science, 11 (2016) 2513 – 2524.

[20] Sheenu Jauhar, Ankitha Goyal, N.Lakshmi, Kailash Chandra, Sonal Singhal, Doping effect of Cr^{3+} ions on the structural, magnetic and electrical properties of Co–Cd ferrites: A study on the redistribution of cations in $CoCd_{0.4}Cr_xFe_{1.6-x}O_4$ $(0.1 \leq x \leq 0.6)$ ferrites, Materials Chemistry Physics,139 (2013) 836- 843. https://doi.org/10.1016/j.matchemphys.2013.02.041

[21] Ferrite Material Science and Technology, Narosa publishing house, New Delhi (1990) 85-105.

[22] D.Ravinder, T.Sheshagiri Rao, Electrical conductivity and Thermo Electric Power studies of Li-Zn ferrites, Crystal Research and Technology 25(1990)1079-1085. https://doi.org/10.1002/crat.2170250918

[23] G. Ravi Kumar, K. Vijaya Kumar, Y. C. Venudhar, Electrical Conductivity and Dielectric Properties of Copper Doped Nickel Ferrites Prepared By Double Sintering Method, International Journal of Modern Engineering Research, 2(2) (2012)177-185.

[24] Manjula, V.R.K. Murthy, J. Shobanadri, Electrical conductivity and thermoelectric power measurements of some lithium–titanium ferrites Journal of Applied Physics 59(1986) 2929-2932. https://doi.org/10.1063/1.336954

[25] Effect of Co Substitution on the dielectric properties of Li-Zn ferrites, Ibetombi, Soibam, Sumita Phanjoubam, Solid State Communications 148 (2008) 399-402. https://doi.org/10.1016/j.ssc.2008.09.029

[26] M. Raghasudha, Ph.D Thesis, Department of Chemistry, Osmania University, August 2013.

[27] M. Pal, P. Brahma, D. Chakravarthy, Magnetic and electrical properties of nickel-zinc ferrites doped with bismuth oxide, Journal of Magnetism and Magnetic Materials, 152 (1996) 370-374. https://doi.org/10.1016/0304-8853(95)00483-1

[28] D. Ravinder, T. Sheshagiri rao, Electrical conductivity and thermoelectric power of lithium-cadmium ferrites, Crystal Research and Technology, 25 (1990) 963-968. https://doi.org/10.1002/crat.2170250820

[29] Vivek Verma, Vibhav Pandey, V.N. Shukla, S. Annapoorni, R.K. Kotnala, Remarkable influence on the dielectric and magnetic properties of lithium ferrite by Ti and Zn substitution, Solid State Communications, 149 (2009) 1726-1730. https://doi.org/10.1016/j.ssc.2009.06.010

[30] Ibetombi Soibam, Sumitra Phanjoubam, H.B. Sharma, H.N.K. Sarma, Effects of Cobalt substitution on the dielectric properties of Li Zn ferrites, Solid State Communications,148 (2008) 399-402. https://doi.org/10.1016/j.ssc.2008.09.029

[31] K.M. Batoo, S. Kumar, C.G. Lee, Alimuddin, Study of dielectric and ac impedance properties of Ti doped Mn ferrites, Current Applied Physics, 9 (2009) 1397-1406. https://doi.org/10.1016/j.cap.2009.03.012

[32] L.L. Hench, J.K. West, Principles of Electronic Ceramics, Wiley, NY, 1990.

[33]   S.C. Watawe, B.D. Sarwade, S.S. Bellad, B.D. Sutar, B.K. Chougale, Microstructure, frequency and temperature-dependent dielectric properties of cobalt-substituted lithium ferrites, Journal of Magnetism and Magnetic Materials, 214 (2000) 55-60. https://doi.org/10.1016/S0304-8853(00)00033-0

[34]   Navneet Singh, Ashish Agarwal, Sujata Sanghi, Study of dielectric and ac impedance properties of Ti doped Mn ferrites, Current Applied Physics, 11 (2011) 783-789. https://doi.org/10.1016/j.cap.2010.11.073

[35]   Navneet Singh, Ashish Agarwal, Sujatha Sanghi, Paramjeet Singh, Synthesis, microstructure, dielectric and magnetic properties of Cu substituted Ni–Li ferrites Journal of Magnetism and Magnetic Materials, 323 (2011) 486-492. https://doi.org/10.1016/j.jmmm.2010.09.053

[36]   C.G. Koop's, On the Dispersion of Resistivity and Dielectric Constant of Some Semiconductors at Audio frequencies, Physical Review, 83 (1951) 121-124. https://doi.org/10.1103/PhysRev.83.121

[37]   Nutan Gupta, S.C. Kashyap, D.C. Dube, Dielectric and Magnetic Properties of Citrate-Route Processed Li-Co Spinel Ferrites, Physica Status Solidi (a), 244 (2007) 2441-2452. https://doi.org/10.1002/pssa.200622146

[38]   I. Soibam, S. Phanjoubam,C. Prakash, Magnetic and Mössbauer studies of Ni substituted Li–Zn ferrite, Journal of Magnetism and Magnetic Materials, 321 (2009) 2779-2782. https://doi.org/10.1016/j.jmmm.2009.04.011

[39]   S.T. Assar, H.F. Abosheisha, Structure and magnetic properties of Co–Ni–Li ferrites synthesized by citrate precursor method, Journal of Magnetism and Magnetic Materials, 324 (2012) 3846-3852. https://doi.org/10.1016/j.jmmm.2012.06.033

[40]   M. Raghasudha, D. Ravinder, P. Veerasomaiah, Magnetic properties of Cr-substituted Co-ferrite nanoparticles synthesized by citrate-gel auto-combustion method, Journal of Nanostruture in Chemistry, 3 (2013) 63-68.

[41]   Q. Zeng, I. Baker, V. McCreary, Zh. Yan, soft ferromagnetism in nano structured mechanical alloying Fe Co based powders, Journal of Magnetism and Magnetic Materials, 318 (2007) 28-38. https://doi.org/10.1016/j.jmmm.2007.04.037

[42]   M. Georgescu, J.L. Viota, M. Klokkenburg, B.H. Erne, D. Vanmaekelbergh, P. Zeijlmans van Emmichoven, Short-range magnetic order in two dimensional cobalt ferrite nano particle assemblies, Physical Review B, 77 (2008) 024423. https://doi.org/10.1103/PhysRevB.77.024423

[43] S.C. Watawe, B.D. Sarwade, S.S. Bellad, B.D. Sutar, B.K. Chaugule, Microstructure and magnetic properties of Li-Co ferrites, Materials Chemistry and Physics, 65 (2000) 173-177. https://doi.org/10.1016/S0254-0584(00)00234-0

Chapter 6

Influence of an Anionic Surfactant Addition on the Structural, Microstructural, Magnetic and Dielectric Properties of Strontium-Copper Hexaferrites

Reshma A. Nandotaria[1,a], Rajshree B. Jotania[1,b], Charanjeet Singh Sandhu[2,c], Sagar E. Shirsath[3,d], Khalid Mujasam Batoo[4,e]

[1]Department of Physics,Gujarat University, Navrangpura, Ahmedabad – 380 009,India

[2]Department of Electronics and Communication Engineering, Lovely Professional University Jalandhar – 144 411, Punjab, India

[3] School of Materials Science & Engineering, The University of New South Wales, Sydney NSW 2052, Australia

[4]King Abdullah Institute for Nanotechnology, King Saud University,Riyadh–11451, Saudi Arabia

[a]nandotaria.riya@gmail.com, [b]rbjotania@gmail.com, [c]rcharanjeet@gmail.com, [d]sagarshirsath@gmail.com, [e]khalid.mujasam@gmail.com

Abstract

The influence of surfactant addition (SDS 1g, 2g, 3g) on structural, microstructural, magnetic and dielectric properties of $Sr_2Cu_2Fe_{12}O_{22}$ hexaferrites, prepared using a coprecipitation method was investigated. XRD analysis show presence of M, Y and α-Fe_2O_3 phases. SEM micrograph of pure sample shows agglomerated spongy grains; while the SDS-1g sample shows plate like hexagonal structure. TEM confirms the formation of nano-particles. The M_s falls from 45 to 31.47 emu/g, while H_c was found to increase from 75 to 500 Oe but M_r/M_s < 0.5. The dielectric properties show normal behavior with respect to frequency.

Keywords

Hexaferrite, XRD, SEM, TEM, Dielectric Properties

Contents

1. Introduction

Ferrites have attracted the attention of the scientific as well as the researcher community for many decades due to their exotic and attractive applications in modern technology. Nowadays, the use of wireless and ultrahigh frequency devices (GHz frequency) such as radar, mobile phones and magnetic refrigeration technologies has been increasing rapidly. Therefore, the usages of these devices led to a serious problem of electromagnetic interference (EMI) and electromagnetic radiation (EMR) with the equipment. This entices the researchers to make new materials, which can overcome EMR and EMI problems and at the same time make them capable for ultrahigh frequency applications. In this regard, various new multifunctional ferrite materials are being tested almost every day [1-3].

Among the various ferrites, Hexaferrites or mixed ferrites were found to be a very important class of magnetic oxides and these are widely used in electronics and electrical devices, because of their low eddy current, high magnetic stability, electromagnetic and microwave absorbance, high Curie temperature, high resistivity, better dielectric properties, low dielectric loss, extraordinary saturation magnetization (M_s), admirable chemical stability, resistance to corrosion and reasonably low synthesis cost [4, 5].

The Y-structure has six stacked layers of S and T blocks and the space group is a rhombohedral crystal - R_3m [7, 8]. The unit cell is made up of three molecular units (3Y). The Y-hexaferrites all have a preferred plane of magnetization perpendicular to the c-axis at room temperature and possesses high permeability due to spin rotation and domain wall motions [9-13]. R.C. Pullar et al. successfully synthesized Co_2Y hexaferrite using novel aqueous sol-gel method [14]. It is reported that Co_2Y show planer anisotropy at room temperature but changes to a cone below − 58 °C, where as Cu_2Y is inclined to have a uniaxial direction of magnetization [9, 12]. K. Taniguchi et al. reported multiferroic/magnetoelectric property of Y-type, mono phased $Ba_2Mg_2Fe_{12}O_{22}$ hexaferrite at cryogenic temperature [8]

Y-hexaferrite is a soft magnetic material and used in electronic communication, microwave devices, electronic components, VHF and UHF [12,15-17]. Y-type hexaferrites are considered as potential materials at relatively lower microwave frequency (< 10 GHz) because of their high permeability.

We have prepared $Sr_2Cu_2Fe_{12}O_{22}$ hexaferritepowder in presence of 1g, 2g and 3g sodium dodecyl sulphate (SDS) using the chemical co-precipitation method. The powder of an anionic surfactant – SDS was mixed in double distilled de-ionized water and used in the synthesis to restrict the growth of the grains. Surfactant coating keeps nucleated legends separate to each other because of static force, as a result surfactant prevents agglomeration of grains and uniformed grains can be obtained. The aim of the present study is to investigate the effect of SDS (1g, 2g, 3g) addition on the structural, morphological, magnetic and dielectric properties of $Sr_2Cu_2Fe_{12}O_{22}$ hexaferrite nanoparticles.

2. Experimental procedure

2.1 Synthesis of strontium-copper hexaferrites

$Sr_2Cu_2Fe_{12}O_{22}$ hexaferrite powder samples were prepared using the chemical co-precipitation technique in the presence of an anionic surfactant- SDS (1g, 2g, 3g). A. R. Grade powder of strontium nitrate $Sr(NO_3)_2$ (Sigma Aldrich, 99.9 % pure), cupric nitrate $Cu(NO_3)_2$ (Sigma Aldrich, 99.9 % pure), ferric nitrate $Fe(NO_3)_3 \cdot 9H_2O$ (Sigma Aldrich, 98 % pure) was used as starting materials. An ammonium hydroxide (25 % w/v NH_4OH) solution was used as precipitating agent and sodium dodecyl sulphate - SDS (1g, 2g, 3g) was used as an anionic surfactant. Stoichiometric amount of strontium nitrate, cupric nitrate, ferric nitrate were mixed with an appropriate amount of double distilled de-ionized water. An appropriate amount (1g/ 2g/ 3g) of surfactant added to the mixed solution and the prepared mixture was stirred at 150 rpm for 2 h, then ammonium hydroxide solution (25 % w/v) was added slowly to the mixture to adjust pH of 11. The mixed solution was stirred for 2 h and was kept at room temperature for 24 h for aging. The obtained precipitates were filtered and then washed with methanol-acetone (1:1) mixture, followed by double distilled water in order to remove impurities. The precipitate was dried at 100 °C for 24 h and then heated at 950 °C for 4 h in a muffle furnace to obtain a final product of $Sr_2Cu_2Fe_{12}O_{22}$ hexaferrite powder. Fig.1 shows the schematic diagram for the preparation of $Sr_2Cu_2Fe_{12}O_{22}$ hexaferrite powder in the presence of SDS (1g, 2g, 3g each) surfactant using chemical co-precipitation technique.

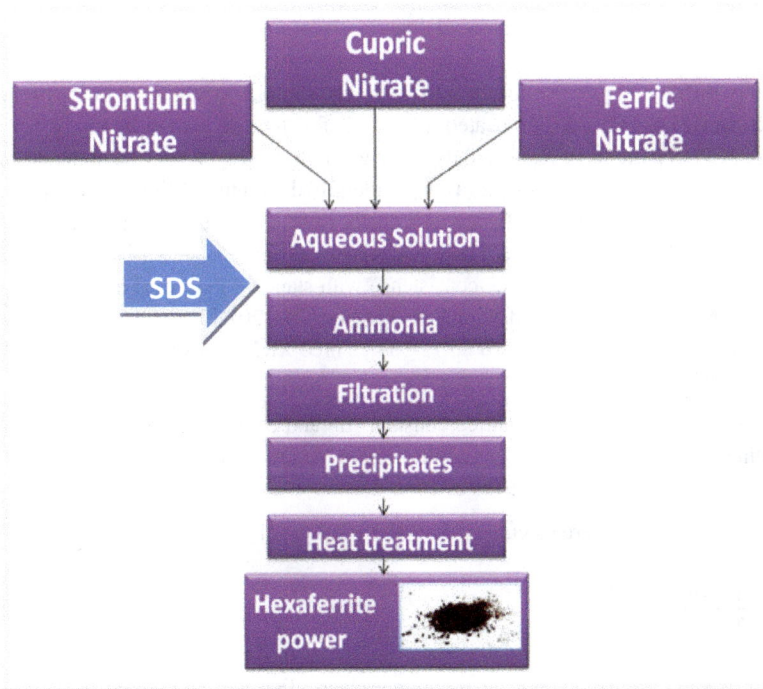

Fig. 1. Schematic diagram for the preparation of $Sr_2Cu_2Fe_{12}O_{22}$ hexaferrite powder.

3.   Characterisation

XRD patterns of $Sr_2Cu_2Fe_{12}O_{22}$ powder samples synthesized in the presence of SDS (1g, 2g, 3g), heated at 950 °C for 4 h. recorded at room temperature on Philips diffractometer (PW 1830) using Cu Kα radiation (λ = 1506 Å) in the range of 2θ = 25-70°. SEM images of all prepared samples were recorded on a Scanning Electron Microscope (SEM- Make Leo/Lica Model Steroecan 440). TEM images of SDS 1g, 3g samples were recorded with a Transmission Electron Microscope (TEM, Tecnai, G2 (square), 20S-Twin, FEI Netherland). The hysteresis loops were recorded using a Vibrating Sample Magnetometer (VSM) on E G & G Princeton Applied Research instrument (Model-4500). The dielectric measurements were carried out at room temperature using an Agilent Precision LCR meter Model no. E4980A in the frequency range of 100 Hz to 2 MHz.

4.   Results and discussion

4.1   XRD analysis

The X-ray diffraction patterns for the $Sr_2Cu_2Fe_{12}O_{22}$ samples synthesized in the presence of surfactant SDS (1g, 2g, 3g) heated at 950° C for 4 h are shown in Fig. 2 (a) and standard JCPDS patterns of Y, M and α-Fe_2O_3 are shown in Fig. 2 (b).The obtained sharp and well defined XRD peaks (Fig. 2 (a)) were identified with their Miller indices [18] and indexed using the Powder-X software. All peaks were compared with standard JCPDS data cards of Y, M and α-Fe_2O_3 [19]. The structural analysis of all samples indicate formation of Y-type (major) phase, consistence with standard PDF card no. 44-0206, M-type (minor) phase standardized with PDF card no. 24-1207 [20] and α-Fe_2O_3 (minor) consistence with standard PDF card no.13-534. Three XRD peaks (minor at 2θ ~27.29°, 44.50° and 62.82°) are unidentified.

The lattice parameters such as lattice constant (a and c), unit cell volume (V_{cell}), and crystalline size (D_{xrd}) of all samples, synthesized in the presence of SDS (1g, 2g, 3g) were calculated using XRD data and listed in Table 1.

The lattice parameters were calculated from the relation [21]

$$\frac{1}{d^2hkl} = \frac{4}{3}\frac{h^2+k^2+l^2}{a^2} + \frac{l^2}{c^2} \tag{1}$$

Where h, k and l are Miller indices, d_{hkl} is interplanar distance. The lattice volume of all the samples was calculated using Eq. (1) [19]

$$V = \frac{\sqrt{3}}{2}a^2c \tag{2}$$

The average crystalline size (D_{xrd}) of the samples was calculated by considering the most intense Bragg peak using the Debye-Scherrer formula [22]

$$D_{xrd} = 0.9 \frac{\lambda}{\beta \cos\theta} \tag{3}$$

Where λ is the X-ray wavelength, θ is the angle of Bragg diffraction and β is the difference between FWHM and the instrumental broadening [23].

The obtained lattice parameters are in agreement with other reported Y-type hexaferrites [24-28] XRD analysis of all the samples confirm formation of mixed phases of Y- M-type (hexagonal crystal structure)and α-Fe_2O_3 (rhombohedral).

The average crystallite size for (D_{xrd}) all the samples varies from ~26 to 38 nm for the samples synthesized in presence of SDS as shown in Table. 1. It seems the average crystallite size of the samples increases in the presence of SDS. The variation of lattice constants and unit cell volume with SDS addition is shown in Fig. 2 (c).

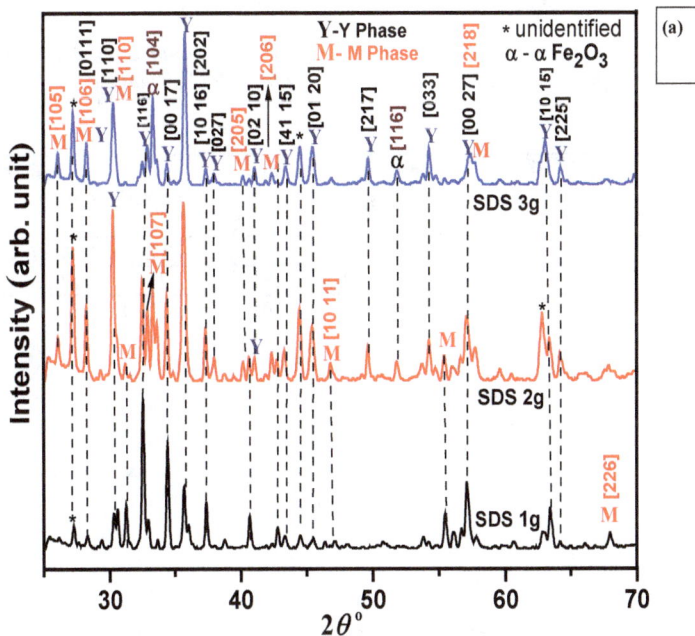

Fig. 2(a). X-ray diffraction patterns of $Sr_2Cu_2Fe_{12}O_{22}$ powder samples synthesized in the presence of SDS (1g, 2g, 3g), heated at 950° C for 4 h.

167

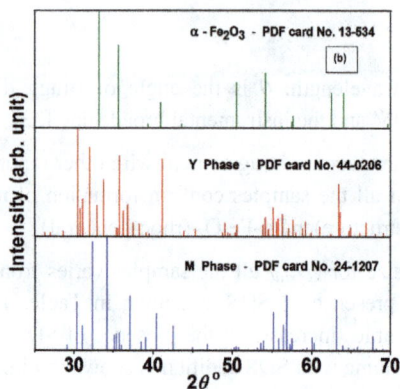

Fig. 2(b). Standard JCPDS patterns of α-Fe_2O_3, M and Y- type phases.

Table 1. Lattice parameters (a and c) and unit cell volume (V_{cell}) for $Sr_2Cu_2Fe_{12}O_{22}$ hexaferrite samples synthesized in the presence of SDS (1g/2g/3g each), heated at 950°C for 4 h.

$Sr_2Cu_2Fe_{12}O_{22}$ sample	Lattice constants		Unit cell volume V_{cell} (A^3)	Average crystallite sizeD_{xrd} (nm)
	a (Å)	c (Å)		
SDS- 1 g	5.850	43.598	1292	37.56 ± 1.18 for [116]
SDS- 2 g	5.852	43.590	1293	25.55 ± 1.28 for [202]
SDS- 3 g	5.885	43.606	1304	31.10 ± 1.56 for [202]

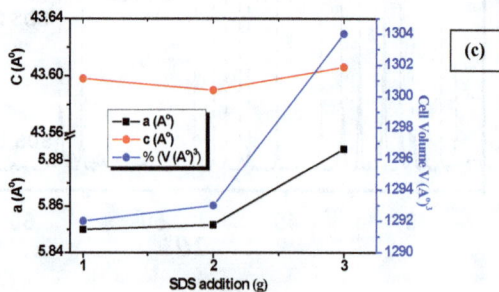

Fig. 2(c). The variation of lattice constants and unit cell volume with SDS addition of $Sr_2Cu_2Fe_{12}O_{22}$ hexaferrite powder samples synthesized in the presence of SDS (1g, 2g, 3g), heated at 950°C for 4 h.

4.2 Microstructure and morphology analysis

The microstructure and surface morphology of $Sr_2Cu_2Fe_{12}O_{22}$ powders were examined by Scanning Electron Microscopy (SEM) and Transmission Electron microscopy (TEM). Fig. 3 (a) shows the SEM images of pure and $Sr_2Cu_2Fe_{12}O_{22}$ hexaferrite samples synthesized in the presence of SDS (1g, 2g ,3g), heated at 950° C for 4 h. It is seen from the SEM images, that the pure sample possesses non-homogeneous agglomerated clusters, while the SDS 1g sample shows thin, stacked hexagonal plates (indicates in-plane preferential crystallization) along with smaller but non-hexagonal clusters. SEM images of 2g, 3g samples show relatively wide, non uniform clusters, form porous structure.

The TEM images of typical $Sr_2Cu_2Fe_{12}O_{22}$ (SDS-1g, 3g) samples are shown in Fig. 3 (b). TEM image of the SDS 1g sample shows clusters of agglomerated nanoparticles (<100 nm) having hexagonal platelet like structure, while the SDS 3g sample indicates spherical and uniform nanoparticles (< 50 nm). It seems the agglomeration is found to decrease [29-31] and the morphology of the grown particles changed drastically when more surfactant amount is used during synthesis of present hexaferrites. One can conclude that the surfactant molecules may serve as an agglomeration inhibitor and led to smaller size, segregated particles, which is caused by the stabilization properties of surfactant.

4.3 Thermomagnetic curves

Thermomagnetic measurements were carried out on prepared Sr-Cu hexaferrite powder samples under low field (10 Oe). The measurements were performed during warming up of the powder samples from 40° C to 550° C. The variation of magnetization as a function of temperature and the first derivative curves of $Sr_2Cu_2Fe_{12}O_{22}$ hexaferrite powder samples are shown in Fig 4 (a-c). All the samples show two transitions; which are associated with Y and M phases. The first transition corresponds to the Curie temperature, at which ferrimagnetic to paramagnetic transition starts to appear. Measurement of the Curie or Néel temperature [32] is often used to identify magnetic minerals in rocks or sediments [33]. These measurements are also used to obtain the size distribution and anisotropy of magnetic nanoparticles [34]. The derivatives of magnetization curves of Fig. 4(a, b) show two transitions at 370° C and 410° C for SDS 1g, 2g samples respectively, which are associated with two hexaferrite phases (Y and M-type) in the present samples [35]; XRD analysis also confirms formation of Y and M phases along with α-Fe_2O_3. The second peak represents magnetic phase transition. The positions of first as well as second peaks are however not fixed in the SDS 3g sample, so it is difficult to find correct values of Curie temperature and magnetic phase transition in the present sample.

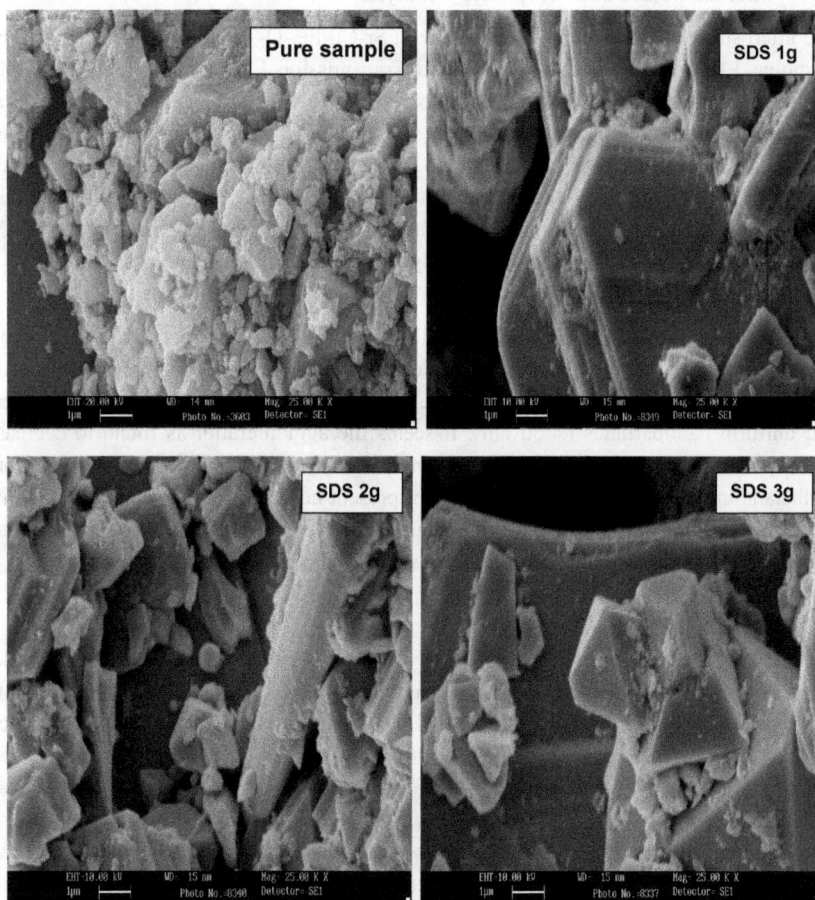

Fig. 3(a). SEM images of Sr₂Cu₂Fe₁₂O₂₂ samples, synthesized without (pure) and with the presence of SDS (1g, 2g, 3g), heated at 950° C for 4 h.

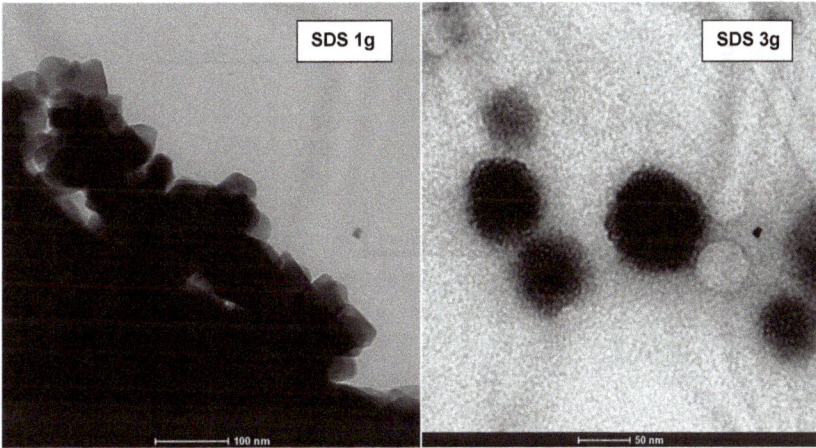

Fig. 3(b). TEM images of typical $Sr_2Cu_2Fe_{12}O_{22}$ samples, synthesized in the presence of SDS (1g and 3g), heated at 950° C for 4 h.

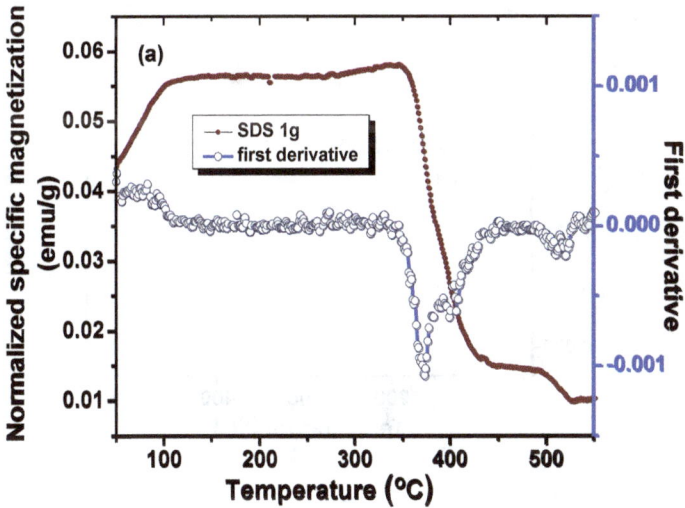

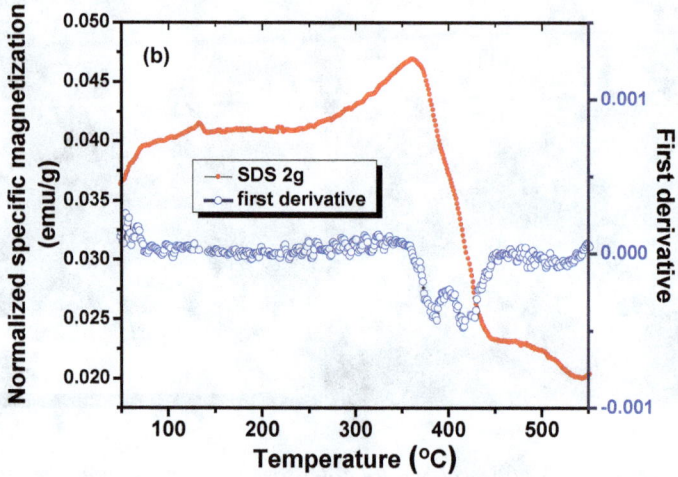

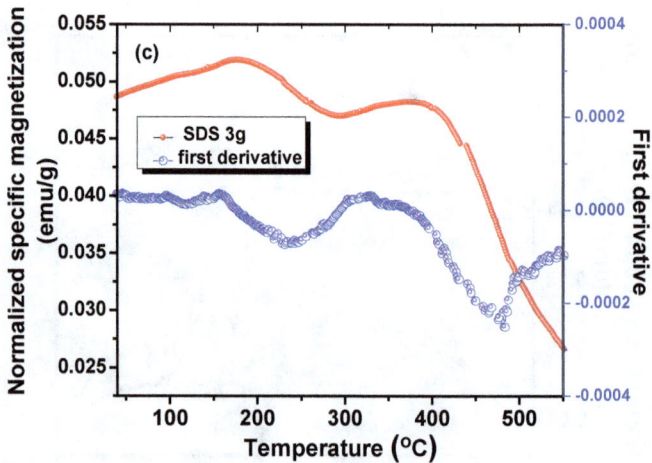

Fig. 4(a-c). Variation of normalized specific magnetization with temperature of the $Sr_2Cu_2Fe_{12}O_{22}$ hexaferrite powder sample synthesized in the presence of SDS (1g, 2g, 3g), heated at 950° C for 4 h.

4.4 Magnetic properties

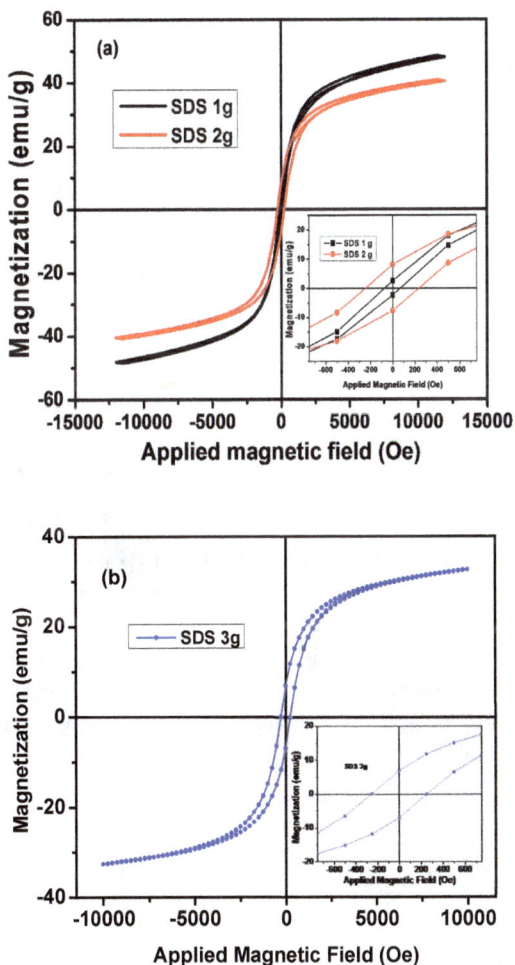

Fig. 5 (a, b). The hysteresis loops of $Sr_2Cu_2Fe_{12}O_{22}$ hexaferrite samples synthesized in presence of SDS (1g, 2 g, 3g), heated at 950° C for 4 h.

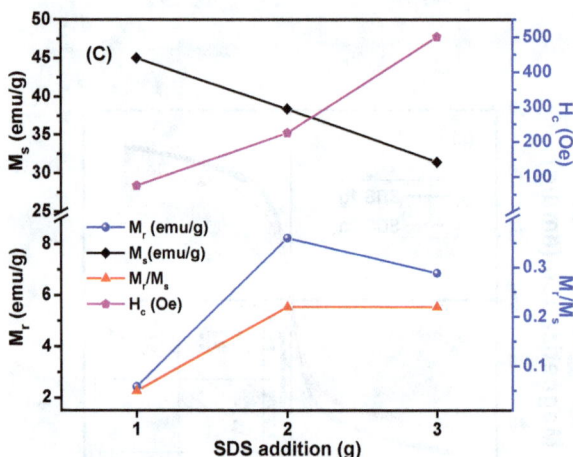

Fig. 5(c). Variation of saturation magnetization (M_s) and remenance magnetization (M_r) corcivity (H_c), and squareness ratio (M_r/M_s) with SDS addition in $Sr_2Cu_2Fe_{12}O_{22}$ hexaferrite samples synthesized in presence of SDS (1g, 2g, 3g), heated at 950° C for 4 h.

Fig. 5 (a, b) shows hysteresis loops for $Sr_2Cu_2Fe_{12}O_{22}$ samples synthesized in the presence of SDS (1g, 2g, 3g), heated at 950° C for 4 h. The hysteresis loop measurements of SDS 1g, 2g samples were carried out at room temperature under an applied field of 12 kOe, while for SDS 3g. sample, hysteresis loop was recorded under an applied field of 10 kOe. The magnetic parameters like saturation Magnetization (M_s), coercivity (H_c), remanent magnetization (M_r) and squareness ratio (M_r/M_s) were calculated from hysteresis loops and listed in Table 2.

The saturation magnetization of the pure $Sr_2Cu_2Fe_{12}O_{22}$ sample has been reported ~60 emu/g [12], but the observed low values of saturation magnetization in prepared samples; may be due to the presence of α-Fe_2O_3 (XRD patterns show clear evidence of mixed phases of Y, M – type, α-Fe_2O_3). Our previous study on $Sr_2Cu_2Fe_{12}O_{22}$ samples synthesized in presence of 1g Tween 80, heated at 950 °C for 4 h shows M_s = 40.404 emu/g, M_r = 13.46 emu/g, H_c = 458.82 and M_r/M_s = 0.30, while 2g Tween 80, heated at 950 °C for 4 h shows M_s = 38.98 emu/g, M_r = 12.446 emu/g, H_c = 457.14 and M_r/M_s = 0.31 [35]. $Sr_2Cu_2Fe_{12}O_{22}$ samples synthesized in presence of 1g, 2g, 3g CTAB, heated at 950 °C for 4 h show variation of M_s from 39.12 to 48.54 emu/g, M_r from 14.81 to 8.01 emu/g, H_c from 429.43 to 233.12 Oe and M_r/M_s from 0.38 to 0.20 [36]. The squareness ratio (M_r/M_s) is an indicative of domain structure and texture of hexaferrite samples. M_r/M_s value of < 0.5 is indicative of presence of multi-domain, randomly oriented assembly of spherical particles; while as lower values are usually associated with larger particles and domain-wall formation, and higher values with texture [35].

All the samples show squareness ratio M_r/M_s < 0.5, indicating formation of the multi-domain structure. For SDS 1g sample M_r/M_s ~ 0.05 and for SDS 2g, 3g samples it is observed 0.22. Such a low value of M_r/M_s = 0.06 is also observed in Zn_2Y prepared at 1200° C with M_s = 32.7 emu/g [37]. In the present case, value of M_s varies between 45.0 to 31.47 emu/g for SDS 1g to 3g samples; which can be attributed to surface effects such as magnetically inactive layer containing spins that are not collinear/aligned with the magnetic field [35, 38-40]. Various theories including surface area, spin canting and sample inhomogenity have been proposed to account for the relatively low saturation magnetization in fine particles [33]. The hysteresis loop of 1g sample shows characteristic of soft ferrite with small value of coercivity (H_c~75 Oe), but the 2g and 3g SDS samples show high value (> 100 Oe) of coercivity.

Table2. Room temperature Magnetic parameters of $Sr_2Cu_2Fe_{12}O_{22}$ hexaferrite powder synthesized in presence of SDS (1g, 2g, 3g) (Coercivity- H_c, Saturation Magnetization-M_s, Remanent Magnetization- M_r measured at 12 kOe for SDS 1g, 2g samples and at 10 kOe for SDS 3g sample)

$Sr_2Cu_2Fe_{12}O_{22}$ sample	M_s (emu/g)	M_r (emu/g)	H_c (Oe)	M_r/M_s
SDS- 1 g	45.0	2.432	75	0.05
SDS- 2 g	38.41	8.237	225	0.22
SDS- 3g	31.47	6.850	500	0.22

4.5   AC conductivity

The frequency response curves of all prepared the samples (Fig. 6 (a)) show normal behavior, initially, as the frequency increases, the variation of AC conductivity is slow from 10 KHz to 100KHz , but there is a rapid increase for the frequency region above 100 KHz to 1MHz. The observed frequency response of the samples at higher frequency values of the applied field is found to agree with the reports of other ferrites [41-43]. The conduction mechanism in ferrites explained on the basis of hopping model [18]. Hopping in ferrites involves the collective jumping of carriers between the similar type, but differently excited metal cations, i.e., $Fe^{2+} \leftrightarrow Fe^{3+}$. The behavior can be explained on the basis that initially as the frequency of the field increases, the charges do not obey the field due to the inertia. However, as the applied field is increased, they start following the field and as such hopping between the two Fe states increases. The increased field liberates, pushes more and more charges, which take part in the conduction mechanism.

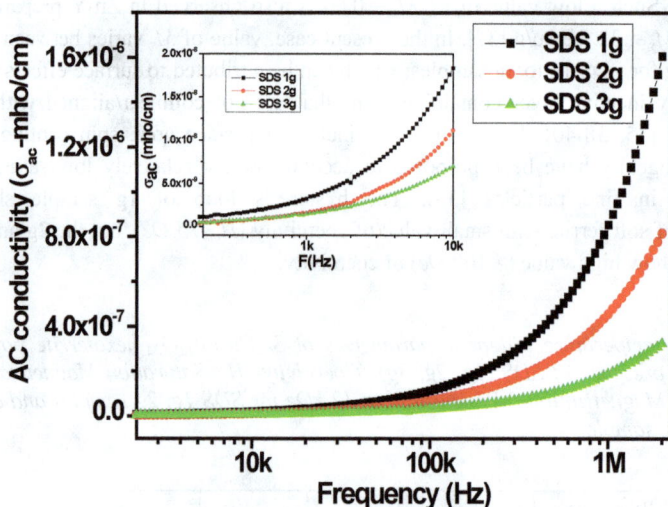

Fig. 6 (a). Variation of AC conductivity (σ_{ac}) with frequency for $Sr_2Cu_2Fe_{12}O_{22}$ hexaferrite powder samples synthesized in presence of SDS (1g, 2g, 3g), heated at 950 °C for 4 h.

4.6 Dielectric response

The dielectric constant of all samples was measured at room temperature in the frequency range of 100 to 2 MHz and result is shown in Fig. 6 (b). All the samples show normal behavior with frequency, it decreases with increasing frequency of an applied field. The behavior can be explained using the Maxwell-Wagner type two layers model [44] in agreement with the Koop's phenomenological theory [45]. Another attributes for variation in the conductivity are (i) grain boundaries are effective in the low frequency region impeding the movement of the

charge carriers which decrease the conductivity (ii) grains are effective in the high frequency region, thus, conductivity increases.

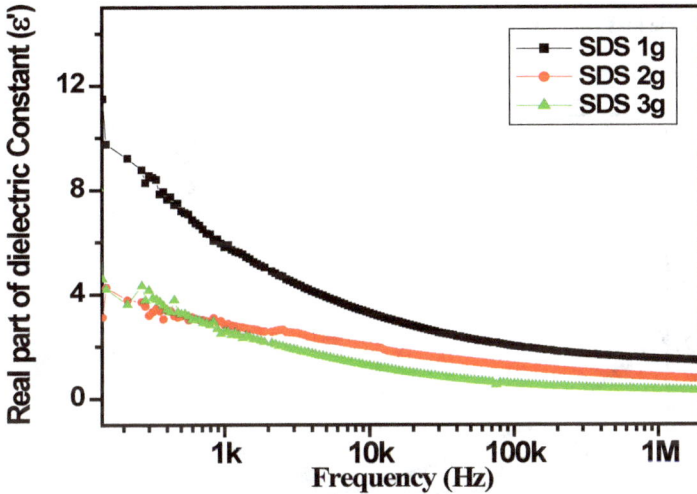

Fig. 6 (b). Variation of real part of dielectric constant (ε') with frequency for $Sr_2Cu_2Fe_{12}O_{22}$ hexaferrite samples, synthesized in presence of SDS (1g, 2g, 3g) and heated at $950° C$ for 4 h.

The hexaferrites possess heterogeneous structure and can be assumed to be made up of two layers; the first layer contains good conducting grains and the second layer consists of highly resistive insulating thin grain boundaries. The electrons reach the insulating grain boundaries through hopping mechanism and accumulate due to resistivity of the material. The accumulation of charges result in the increase of interfacial polarization under the influence of the increased frequency of an applied field, and hence higher value of dielectric constant is observed at low frequency. The dielectric constant naturally decreases with the increase of an applied field reversal frequency due to the fact that the charge carrier's cannot follow the changes of an applied field at high frequency because in inhomogeneous dielectric structure requires a finite time to align themselves along with the direction of an applied field. In the low frequency region, the hopping of charges between the two Fe sites, i.e., Fe^{2+} and Fe^{3+} synchronize with an applied field, but in the high frequency region, this synchronization is no longer able to be maintained and lags behind an applied field, and hence one see a frequency independent like response [44].

4.7 Dielectric loss tangent

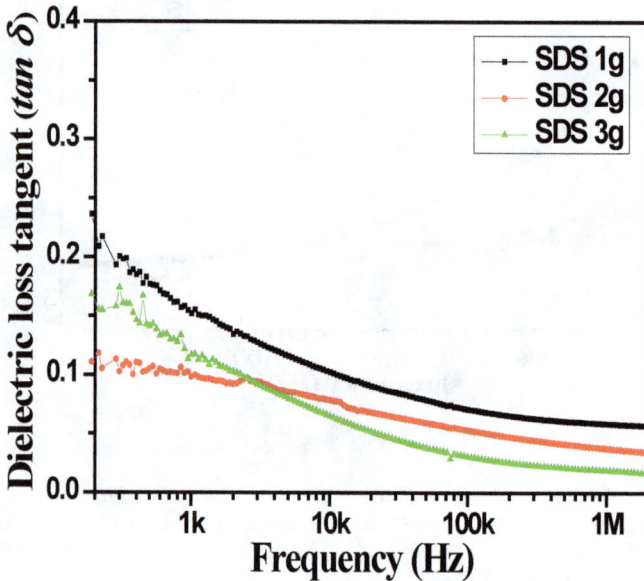

Fig. 6(c). Variation of dielectric loss tangent (tan δ) with frequency for $Sr_2Cu_2Fe_{12}O_{22}$ samples, synthesized in presence of SDS (1g, 2g, 3g) heated at 950° C for 4 h.

The dielectric loss tangent depends on many factors such as stoichiometry ratio, number of Fe^{2+} present, structural homogeneity, and preparation conditions such as heat treatment, heating temperature and time of sintering etc. It represents the energy dissipation in a dielectric material and hence determines applicability of a material in high frequency regions. Fig.6(c) shows variation of dielectric loss tangent with frequency at room temperature in the frequency range 100 to 2 MHz. The dielectric loss tangent behavior shows a similar trend as observed in dielectric constant with frequency; it decreases with increasing frequency of an applied field. The behavior can be explained on the basis, the hopping frequency of the charge between Fe^{3+} and Fe^{2+} cannot follow the frequency of the applied alternating electric field beyond certain frequency limit. There is a strong correlation between the conduction mechanism in ferrites and the dielectric behaviour. As the frequency of the field increases the resistivity (which predominantly comes through the grain boundaries) increases, which in turn decreases the charge exchange mechanism and hence dielectric loss factor increases [18, 44-46].

Conclusions

Strontium-copper hexaferrites with chemical composition $Sr_2Cu_2Fe_{12}O_{22}$ were successfully synthesized in presence of an anionic surfactant SDS (1g, 2g, 3g). Influence of SDS addition on the structural, microstructural, magnetic and dielectric properties was investigated using XRD, SEM, VSM, low field AC susceptibility and dielectric measurements. XRD analysis confirms the formation of mixed phases of Y, M and α-Fe_2O_3. SDS addition has improved the surface morphology of prepared materials. Saturation magnetization is found to maximum 45.0 emu/g and minimum 31.47 emu/g. Magnetic analysis confirms formation of multi domain structure. The dielectric behavior of all prepared samples was found to be normal and is explained using the Maxwell-Wagner's model.

Acknowledgements

This work was carried out under DRS-SAP, phase-I program of UGC, New Delhi, India (letter no. UGC-F.530/10/DRS/2010 (SAP-I)), and DST-FIST-Level-I, India. One of the authors (R. A. Nandotaria) acknowledges to the University Grant Commission, New Delhi, India (Grant no F1-17.1/2012-13/RGNF-2012-13-SC-GUJ-34700/9SA-III/website) for providing Rajiv Gandhi National Fellowship and K.M. Batoo is thankful to the Deanship of Scientific Research at King Saud University, Riyadh, Saudi Arabia for its funding through the Research Group Project No. RG-1437-030.

References

[1]   Z.H. Yang, Z.W. Li, L. Liu, L.B. Kong. Enhanced microwave magnetic and attenuation properties of composites with freestanding spinel ferrite thick films as fillers, Journal of Magnetism and Magnetic Materials, 324 (19) (2012) 3144-3148.

[2]   R.S. Alam, M. Moradi, M. Rosta, L. Y. Bai. Structural, magnetic and microwave absorption properties of doped Ba-hexaferrite nanoparticles synthesized by co-precipitation method, Journal of Magnetism and Magnetic Materials, 381 (2015) 1-9.

[3]   A. Ohlan, K.Singh, A.Chandra, S.K.Dhawan, Microwave absorption behavior of core shell structured poly (4-ethylenedioxy Thiophene) barium ferrite nanocomposites, ACS Applied Materials and Interfaces, 2 (3) (2010) 927-933.

[4]   K. Kamishima, N. Hosaka, K. Kakizaki, N. Hiratsuka, Crystallographic and magnetic properties of Cu_2X, Co_2X, and Ni_2X hexaferrites, Journal of Applied Physics, 109 (1) (2011) 013904-5.

[5]   Z. Zhang, X. Liu, X.Wang, Y. Wu, R. Li, Effect of Nd-Co substitution on magnetic and microwave absorption properties of $SrFe_{12}O_{19}$ hexaferrites, Journal of Alloys and Compounds, 525 (2012) 114-119.

[6]   Z. F. Zi, Y. P. Sun, X. B. Zhu, Z. R. Yang, J. M. Dai, W. H. Song, Structural and magnetic properties of $SrFe_{12}O_{19}$ hexaferrite synthesized by a modified chemical co-precipitation method, Journal of Magnetism and Magnetic Materials, 320 (21) (2008) 2746-2751.

[7]   P.B. Braune, The crystal structures of a new group of ferromagnetic compounds, Philips research reports,12 (1957) 491-548.

[8]   K. Taniguchi, N. Abel, S. Ohtani, H. Umetsul and T. Arima, Ferroelectric polarization reversal by a magnetic field in multiferroic Y-type hexaferrite $Ba_2Mg_2Fe_{12}O_{22}$,Applied Physics Express, 1 (2008) 031301-3.

[9]   J. L. Snoek, Dispersion and absorption in magnetic ferrites at frequencies above one Mc/s Physica, 14 (4) (1948) 207-217.

[10]   J. Smit, H P J Wijn, Ferrites, Cleaver-Hume Press, London, UK, (1959).

[11]   R.C. Pullar, Hexagonal ferrites: a review of the synthesis, properties and applications of hexaferrite ceramics, Progress in Materials Science, 57 (2012) 1191-1334.

[12]   R. B. Jotania, H. S. Virk, Y-Type hexaferrites: structural, dielectric and magnetic properties, Solid State Phenomena, 189 (2012) 209-232.

[13]   L. Bolin Hu, Zhaohui Chen, Zhijuan Su, Xian Wang, Andrew Daigle, Parisa Andalib, Jason Wolf, Michael E. McHenry, Yajie Chen and Vincent G. HarrisNanoscale-driven crystal growth of hexaferrite heterostructures for magnetoelectric tuning of microwave semiconductor integrated devices, ACS Nano, 8 (11) (2014) 11172–11180

[14]   R.C. Pullar, M.D. Taylor, A. K. Bhattacharya, Novel aqueous sol–gel preparation and characterization of barium M ferrite, $BaFe_{12}O_{19}$ fibers, Journal of Materials Science, 32 (1997) 349-352.

[15]   A. Morel, J. M. LeBreton, J. Kreisel, G. Wiesinger, F. Kools and P. Tenaud, Sublattice occupation in $Sr_{1-y}La_xFe_{12-y}CoO_{19}$ hexagonal ferrite analyzed by Mossbauer and Raman spectroscopy, Journal of Magnetism and Magnetic Materials. 1405 (2002) 242-245.

[16]   Y. Bai, J. Zhou, Z. Gui and L. Li, Magnetic properties of non-stoichiometric Y-type hexaferrite, Journal of Magnetism and Magnetic Materials. 250 (2002) 364-369.

[17]   M. Obol, X. Zuo and C. Vittoria, Oriented Y-type hexaferrites for ferrite device, Journal of Applied Physics, 91 (2006) 7616-7618.

[18]   J.C. Maxwell, Electricity and Magnetism, Oxford University Press, New York, (1954) 328.

[19]   J. J. Went, G.W Rathenau, E.W Gorter, G.W Oosterhout, Hexagonal iron oxide compounds permanent-magnet materials, Philips technical review, 13(1952) 194-208.

[20]   W. D Kingery, H.K Bowen, D.R Uhlmann, Introduction to Ceramics, second ed, Wiley, New York, 1976.

[21]   A. R. West, Solid State Chemistry and its applications, John Wiley & Sons, New York, 1984.

[22]   W.H. Bragg, W. L. Bragg, The reflection of X-rays by crystals, Proceedings of the Royal Society of London. Series A, 88 (605) (1913) 428-438.

[23]   B. D. Cullity, Elements of X-ray Diffraction, Addison-Wesley Publishing company, Printed at USA, 1956.

[24]   U. A. Barkat, S. Ahmed, Y. Huang, Catalytic decomposition of N_2O on cobalt substituted barium hexaferrites, Chinese Journal of Catalysis, 34 (7) (2013) 1357-1362.

[25]   S. Bierlich, J. Töpfer, Zn-and Cu-substituted Co_2Y hexagonal ferrites: Sintering behavior and permeability, Journal of Magnetism and Magnetic Materials, 324 (10) (2012) 1804-1808

[26]   T. Koutzarova, S. Kolev, I. Nedkov, K. Krezhov, D. Kovacheva, B. Blagoev, C. Ghelev, C. Henrist, R. Cloots, A. Zaleski, Magnetic properties of nanosized $Ba_2Mg_2Fe_{12}O_{22}$powders obtained by auto-combustion, Journal of superconductivity and novel magnetism, 25 (8) (2012) 2631-2635.

[27] M. Costa, .P. Júnior, A. Sombra, Dielectric and impedance properties' studies of the of lead doped (PbO)-Co_2Y type hexaferrite ($Ba_2Co_2Fe_{12}O_{22}$ (Co2Y)), Materials Chemistry and Physics, 123 (1) (2010) 35-39.

[28] Sami H. Mahmood, Muna D. Zaqsaw,Osama E. Mohsen, Ahmad Awadallah, Ibrahim Bsoul, Mufeed Awawdeh, Qassem I. Mohaidat, Modification of the magnetic properties of Co_2Y hexaferrites by divalent and trivalent metal substitutions, Solid State Phenomena, 241 (2016) 93-125

[29] K Nejati, R Zabihi, Preparation and magnetic properties of nanosize Nickel ferrite particles using hydrothermal method, Chemistry Central Journal, 6 (2012) 23-29.

[30] K. K. Patankar, S.S. Joshi, B. K. Chougule, Dielectric behaviour in magmetoelectric composites, Physics Letters A, 346 (2005) 337-341.

[31] A. A. Birajdar, S. E. Shirsath, R. H. Kadam, S. M. Patange, D. R. Mane, A. R. Shitre, Frequency and temperature dependent electrical properties of $Ni_{0.7}Zn_{0.3}Cr_xFe_{2x}O_4$, Ceramic International, 38 (2012) 2963-2970.

[32] I .G. Austin, N. F. Mott, Polarans in crystalline and noncrystalline materials, Advances in Physics, 18 (1969) 41-102.

[33] M. Hanesch, H. Stanjek, N. Petersen, Thermogravimatric measurements of soil iron minerals: the role of organic carbon, Geophysical Journal, 165(1) (2006) 53-61.

[34] R.S. Di Pietro, H.G. Johnson, S. P. Bennett, T. J. Nummy, L. H. Lewis, D. Heiman, Determining magnetic nanoparticle size distributions from thermomagnetic measurements, Applied Physics Letter, 96 (22) (2010) 222506 (1-4).

[35] R. A. Nandotaria, C. C. Chauhan, R. B. Jotania, Effect of non-ionic surfactant concentration on microstructure, magnetic and dielectric properties of strontium copper hexaferrite powder, Solid State Phenomena, 232 (2015) 93-110.

[36] Reshma A. Nandotaria, Ph. D. thesis, Department of Physics, Gujarat University, 2015.

[37] T.O. Kim, S. J. Kim, P. Grohs, D. Bonnenberg, K.A. Hempel, Ferrites, proc ICF6, Tokyo and Kyoto, (1992) 75.

[38] M. Yamaura, R.L. Camilo, L.C. Sampaio, M. A. Macedo, M. Nakamura, H. E. Toma, Preparation and characterization of triethoxysilane-coated magnetite nanoparticles, Journal of Magnetism and Magnetic Materials, 279 (2004) 210-217.

[39] S. Tumanski, Handbook of magnetic measurements. Boca Raton, 2011.

[40] R. A. Nandotaria, R. B. Jotania, A solvent free synthesis and characterization of Y-type hexaferrite particles in presence of a cationic surfactant, Journal of International Academy of Physical Sciences, 17 (2013) 81-86.

[41] E. Melagriyappa, H.S. Jayanna, B.K. Chougule, Dielectric behavior and ac electrical conductivity study of Sm^{3+} substituted Mg–Zn ferrites, Materials Chemistry and Physics,112 (2008) 68-73.

[42] Q. A. M.bo El Ata, M. K. Elimr, S. M. Attia, D. El Kony, A. Al-Hammadi, Studies of AC electrical conductivity and initial magnetic permeability of rare-earth-substituted Li–Co ferrites, Journal of Magnetism and Magnetic Materials, 297 (2006) 33-43.

[43] L.T. Rabinkin, Z.I. Novikova, Ferrites, Minsk: Influence of Cr^{3+} ion on the dielectric properties of nano crystalline Mg-ferrites synthesized by citrate-gel method. Akademii Nauk, USSR, 12 (1960) 146.

[44] K.W. Wagner, The distribution of relaxation times in typical dielectrics, American Journal of Physics, 40 (1973) 817-819

[45] C.G. Koops, On the dispersion of resistivity and dielectric constant of some semiconductors at audio frequencies, Physical Review A, 83 (1951) 121-124.

[46] K.M. Batoo, S. Kumar, Synthesisation, electrical and magnetic properties of Al doped nano ferrite particles, International Journal of Nanoparticles, 2 (2009) 416-422.

Chapter 7

Magnetic Nanoparticles: Fabrication, Properties and Applications

B. Parvatheeswara Rao[1,a], Mohamed Abbas[2,b]

[1]Physics Department, Andhra University, Visakhapatnam530003, India

[2]National Research Centre, Egypt & Institute of Coal Chemistry, China

[a]bprao250@gmail.com, [b]mohamed_abbas83@yahoo.com

Abstract

Magnetic nanoparticles are being widely used as potential materials in several applications, namely, consumer electronics, automobile and biomedical fields. Magnetic nanoparticles compacted in an insulating matrix are shown to be excellent materials for miniaturized high frequency electronic devices. Substituted cobalt ferrites and their self composites are increasingly found suitable for non-contact torque sensors. Functionalized magnetic nanoparticles and core-shell nanostructures would prove beneficial for biomedical applications. In this chapter, some simple preparation routes and common characterization techniques for the study of magnetic nanoparticles are covered. Further, recent advances and future prospects in all three application areas mentioned above are briefly discussed.

Keywords

Magnetic Nanoparticles, Synthesis, Ferrites, Torque Sensor, Biomedical Applications

Contents

1. Introduction

Recently, many studies on magnetic nanoparticles ensure that they have immense potential for several applications such as high frequency consumer electronic devices, magnetostrictive sensors and actuators, and biomedical applications like hyperthermia, targeted drug delivery and contrast imaging [1-3]. Besides, magnetic nanoparticles are also widely used for catalytic and environmental protection applications, particularly in the treatment of waste water by removing pollutant organic or inorganic compounds, and methylene blue dye degradation processes [4,5]. Among various magnetic nanoparticles for biomedical and catalytic needs, Fe_3O_4 or magnetite is considered as the most promising kind of magnetic oxide material due to its excellent magnetic properties, good stability and less toxicity apart from exhibiting hydrophilic properties. Further, a nonmagnetic surface coating to the Fe_3O_4 nanoparticles is reported to help in offering an inert shell layer with increased biocompatibility thus enabling the core magnetite nanoparticles not only to survive in vivo but also to work well in specific targeting [6]. Though there are many options to use metal oxides, noble metals and polymer materials as coating materials, silica is considered very promising as an oxide coating material. The silica shell on the surface of the magnetite nanoparticles not only helps in enhancing their biocompatibility, hydrophilicity and stability against degradation but also facilitates easy surface modification due to the availability of abundant silanol groups (-SiOH). This includes strong surface functionalization with amine, thiol and carboxyl groups, and consequently the resultant functionalized nanoparticles become a good choice for biolabelling, drug delivery and targeting applications [7,8]. Furthermore, silica-coated magnetic nanoparticles are also useful for catalytic activity, especially in the conversion of syngas (CO-H_2 mixtures) into a wide range of long chain hydrocarbons and oxygenates via the Fischer-Tropsch [9].

Similarly, ferrite nanoparticles, which are another class of magnetic oxide materials possessing useful exchange coupled magnetic properties along with high resistivities, find themselves very interesting from a technological view point as they are increasingly being used in large quantities as core materials for transformers, inductors, and other sensing and actuating elements. Moreover, inexpensive raw materials, ease in synthesis and excellent magnetic performance make the ferrite nanoparticles attractive over their metal magnetic counterparts for these applications [10]. Besides, the recent up-surge in the study of ferrites at nano-scales for power applications is mainly driven by the assumption of an enhanced structural and magnetic performance over their ceramic counterparts particularly in dealing with the limit due to domain wall resonance [11]. When the size of the ferrite nanoparticle is smaller than the critical size for multi-domain formation, the particle exists in a single domain state and obviously the domain wall resonance is avoided; thus the material can work at higher frequencies.

On the other hand, another class of magnetic materials which are known to exhibit strong magnetostrictive properties are also in widespread use as sensing and actuating elements in many automotive applications. Magnetostrictive material, in general, would change its dimensions slightly when it is placed in the magnetic field due to the Joule effect. Analogous to the tensile or compressive strain produced in the Joule effect, there is a shear strain response to the magnetic field as per the Wiedemann effect. Besides, there is also an inverse effect named after Villari that there arises a change in magnetic permeability in response to an applied stress, which is also referred to as a magnetostrictive effect or a magnetomechanical effect. For the most transducer or sensor applications, maximum force or movement is desired as output and thus the Joule effect or the Villari effect are the most useful in technology.

The advantage of magnetostrictive actuators or sensors over other types is that their driving voltages can be very low. The most commonly used magnetostrictive material in actuation applications was Terfenol-D and some of the applications are sonar transducer, linear motors and rotational motors [12,13]. And, the most common sensors are torque sensor, position sensor and force sensor. Though the Terfenol-D and other intermetallic rare earth compounds such as, Samfenol-E and Galfenol were known as good magnetostrictive materials, they however simultaneously suffer from low anisotropy and high conductivity as drawbacks resulting in limitations in their use. Recently, cobalt ferrites are found as potential alternatives to existing materials competing for use in automotive and other torque sensing applications as they possess good magnetostrictive properties along with high Curie temperature, high resistivity and moderate magnetization. Though the cobalt ferrite compounds have shown promising results in improving stress sensitivity at lower magnetic field strengths, they are also in need of reducing the magnetomechanical hysteresis which can be achieved by employing magnetic annealing as well as chemical modifications. Of late, cobalt ferrite based self composites comprising the mixtures of ferrite particles of different sizes spreading in the ranges from micron to a few nano meters prepared by both conventional solid state methods and wet preparation nano routes are being widely investigated as they have immense potential to cater the needs of non-contact torque and force sensors in several fields [14,15].

Thus the objective of this work is to briefly review the recent developments in the research of magnetic/ferrite nanoparticles using various preparation methodologies and their characterizations, and also to highlight the application potential of these materials for uses in the fields of biomedical, high frequency electronic devices, and sensing and actuating applications.

2.   Synthesis and characterization

There are a large number of methods available in the literature for synthesis of magnetic nanoparticles in different size regimes. Some of the widely used methods for preparation of nanoparticles using chemical processes include hydrothermal, sol-gel autocombustion, polyol, sonochemical, thermal decomposition, etc. Each of these methods were inherited with certain advantages such as ease of synthesis, use of inexpensive raw materials, production of homogenous and uniformly sized particles, etc., some of them are also showing some drawbacks as well including long reaction times and requiring multi-step procedures. For example, the microemulsion and the alkaline hydrolysis of tetraethyl orthosilicate (known as the Stober method) approaches have emerged recently for fabrication of core-shell nanoparticles [16,17]. Though this method is capable of producing nanoparticles with complete silica coating on the surfaces, it not only requires long reaction times but also involves a two-step procedure, in which the first step is for preparation of core magnetic nanoparticles of about 10 nm in size and the second step is for coating a shell of a few nanometers on the surface of the synthesized particles to obtain them as core-shell nanoparticles, and thus obviously it amounts to high costs in execution. Furthermore, some of these methods may require to undergo phase transitions from hydrophobic to hydrophilic or vice versa to be suitable for surface coating with silica.

However, synthesized magnetic nanoparticles should obviously be subjected to a series of characterization procedures with regards to their crystal structure, chemical bonds, nano dimensions, dispersity, magnetic strength, etc. before making any structural modifications and carrying out measurements to ensure that the materials prepared are exactly the ones intended to be studied. The crystal structures of the synthesized magnetic nanoparticles are normally analyzed by X-ray powder diffraction technique (XRD) while the size and morphology of the nanoparticles are characterized using transmission electron microscopy (TEM). Energy dispersive X-ray spectrometer (EDS) gives information related to chemical composition and the Fourier transform infrared (FTIR) spectroscopic data helps to interpret the chemical bonds as well as the traces of surface coating on the nanoparticles. Vibrating sample magnetometer (VSM) provides data related to the magnetic strength of the materials when the synthesized nanoparticles are subjected to external magnetic fields of measurable ranges.

Some of the widely used methods for synthesis as well as characterization of the magnetic nanoparticles relevant to the application areas mentioned in the previous section are briefly given below.

2.1 Synthesis methods

The properties of magnetic nanoparticles strongly depend on the chemical composition as well as the method of synthesis. However, in the quest for finding useful materials with desired characteristics, it has been a common practice to employ either by selecting a specific composition and investigating the same in different synthesis routes or by fixing a specific method for synthesis and then investigating different compositions of a system using that method. The choice of investigation however depends largely upon the specific requirements for an application. Nevertheless, considering the vast potential for magnetic nanoparticles to be used as elements in application systems, it is felt necessary by the scientific community to carry out more and more investigations on magnetic nanoparticles. A study of this kind helps not only in obtaining required materials with desired structural and magnetic characteristics for applications mentioned above but also in understanding the mechanisms responsible for such property transformations in their evolution.

2.1.1 One-pot polyol process

Synthesis procedure in this method involves dissolution of required quantities of $FeCl_2$. $4H_2O$ in PEG using magnetic stirrer in a three–neck round bottomed flask equipped with condenser, magnetic stirrer, thermometer and heating system as reported by Abbas et al. [18]. They adjusted the pH of the solution in between 10-11 by adding NaOH and gradually increased the temperature of the solution to 200° C for 30 minutes. Then, 1 mL of tetraethyl orthosilicate (TEOS) was injected into the solution and further increased the temperature of the PEG-metal salts solution to 300 °C while stirring continuously using a magnetic stirrer, and refluxed at this temperature for 2 h. On completion of the soaking, the heating system was switched off and allowed the solution to cool naturally down to room temperature. The resultant precipitate was then collected using a magnet, washed several times using ethanol and water and then it was dried in a vacuum oven to obtain ultrafine Fe_3O_4/SiO_2 nanoparticles. Schematic diagram for the reaction synthesis of the above nanoparticles was shown in Fig. 1A.

2.1.2 Modified stober method

A- Synthesis and coating of (Fe3O4/SiO2) by single polyol reaction

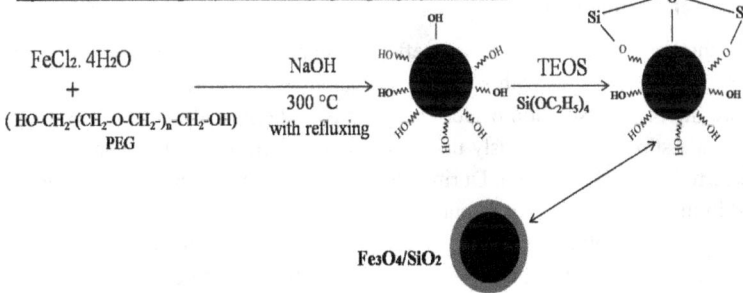

B- Coating (Fe3O4/SiO2) by modified Stober method

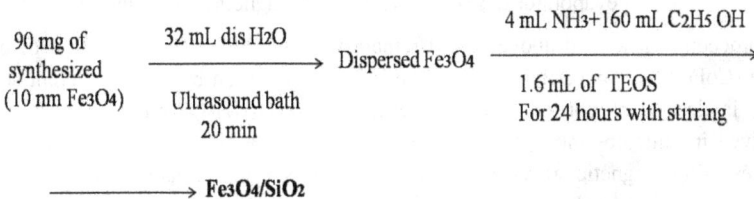

Fig. 1. Schematic diagrams for the synthesis of Fe_3O_4/SiO_2 core-shell nanoparticles using (A) new Polyol method and (B) modified Stober method (Courtesy: Abbas et al. [19]).

189

Synthesis of seed hydrophilic high moment magnetite nanoparticles of about 10 nm in size is a challenging task. Often they might be required to be coated on the surface for additional effects so as to be used in bio or catalytic applications, which require added chemical procedures during synthesis. Abbas et al. [19] in their work used a modified Stober method for the synthesis of seed as well as Fe_3O_4/SiO_2 core-shell magnetite nanoparticles in two separated batches. The first batch was made using a rather simple procedure for preparation of seed magnetite nanoparticles, and the second batch of seed nanoparticles was used for silica coating on them by modified Stober method. This coating was performed by the hydrolysis of Tetraethyl orthosilicate (TEOS) in the presence of magnetite nanoparticles [18] by a series of chemical treatments so as to finally obtain ultrafine Fe_3O_4/SiO_2 core-shell nanoparticles [19]. Schematic diagram for the reaction synthesis of the above nanoparticles was shown in Fig. 1B.

2.1.3 Sonochemical method

There is another method which is very effective in producing magnetite nanoparticles in cubic and spherical shapes with size scales of about 40 nm [20]. In this case, the typical procedure includes dissolution of desired quantity of $FeSO_4$. $7H_2O$ in distilled water for 10 min while stirring continuously using magnetic stirrer, and then the solution needs to be transferred to ultrasonication. During the reaction, an appropriate quantity of NaOH is injected in the reaction after 15 minutes from the start of ultrasonication. Transformation of the colour of the solution from initial blue to black is considered as an indicator for the formation of the magnetite phase, and further addition of some titanium isopropoxide (TIPP) to this solution while continuing the sonication process for one more hour helps in completion of the reaction. The obtained mixture was finally washed and sonicated for five times in water and ethanol before collecting the precipitate using a magnet and drying in a vacuum evaporator to get the well shaped magnetite nanocubes or spheres.

The procedure described above would remain the same even for fabrication of cobalt ferrite ($CoFe_2O_4$) nanoparticles but with a little modification by using different solvent media in desired proportions [18]. For example, $FeSO_4.7H_2O$ and $CoSO_4.7H_2O$ were dissolved in suitable amounts of distilled water (aqueous medium) in this case for 10 minutes using magnetic stirrer, and then sonicated using an ultrasonic processor for 70 minutes to obtain sample S1 with specific size and features as described in Fig. 2. Then, the authors adopted the same procedure twice again to obtain $CoFe_2O_4$ nanoparticles by changing the solvent medium to ethanol (in this case, the sample is S2), and to a mixed solution of water/ethanol in 1:1 volume ratio (here, it is S3) in place of distilled water. Further, they have also made an attempt to understand the role of the stabilizing agent in the synthesis of $CoFe_2O_4$ nanoparticles by adding little quantity of polyvinyl pyrolidone (PVP) to the aqueous medium.

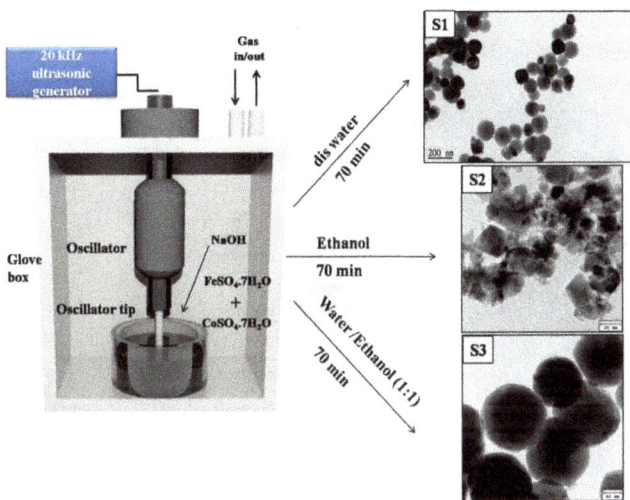

Fig. 2. Ultrasonic generator and the synthesized CoFe₂O₄ nanoparticles in different solvent ratios. (Courtesy: Abbas et al. [20]).

2.1.4 Sol-gel auto-combustion method

Ferrite nanoparticles are reported to behave much different in comparison to their bulk counterparts in terms of their physicochemical properties such as chemical stability, magnetic properties, electrical resistivity, etc. Recently, several groups have synthesized ferrite nanoparticles using wet chemical methods such as coprecipitation [21], citrate gel [22], etc. Though these methods are proved to be good in the control of size and dispersion, some of them also suffer from certain disadvantages too such as low yield, expensive precursors, high pH sensitivity, long synthesis times, etc. However, sol-gel auto combustion process is known to be a simple method to prepare nano scale ferrite powders with inexpensive precursors in relatively large yields [23].

Sol-gel auto/flash combustion process is a modified form of sol-gel process. It may be often called as self propagating high temperature synthesis (SHS) or combustion synthesis (CS) as well. The combustion method was developed by Patil et al. in 1993 [24]. To obtain nanosized powder of ceramic oxides i.e. magnetic, semi conducting oxides, oxides for catalysis, sensors etc., combustion method is one of the best choices. It is simple, rapid and low cost method to obtain nanosized powder of transition metal oxides. The combustion can be carried out in condensed phase, solution phase and gaseous phase depending up on the reactants used. Fuel is an essential ingredient of each route. It is used to activate the combustion/ignition process.

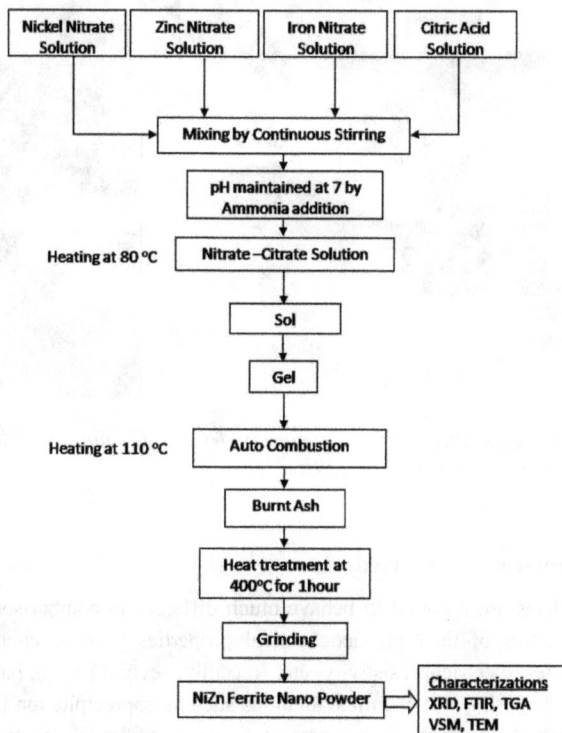

Fig. 3. Flow chart for sol-gel auto combustion synthesis of Ni-Zn ferrite nanoparticles.

Among these routes, solution combustion synthesis is very simple and versatile method to obtain nanosized oxide powder. In this process, particle size varies from 20-60 nm depending upon synthesis conditions. The high yield and compositional homogenization of synthesized product are remarkable advantages of solution combustion synthesis. The exothermicity of ignition/combustion reaction is controlled by the type of fuel and fuel to oxidant ratio. Citric acid, tartaric acid, urea, glycine, hydrazine etc. are the most commonly used fuels.

In the solution combustion process, aqueous solution of desired metal salts is mixed with suitable organic fuel and a strong base is added to adjust pH and then evaporated at 70-80 °C to get dry gel. This dried gel is further heated until ignition started around 100°C to 300 °C. Upon ignition, the dried gel burns in a self-propagating combustion manner until all gels are completely burnt out and converted into fine fluffy powder. The reaction is self-propagating and is able to sustain this from 5 to 10 seconds typically to form the desired product. This technique produces a homogenous product in a short amount of time without the use of expensive high temperature furnaces. A combustion synthesis reaction can be influenced by parameters such as the type of fuel, the de-ionized water content of the precursor mixture and the ignition temperature. The fuel and amount of de-ionized water also plays a critical role in influencing the reaction. A way to control the flame temperature of the reaction is by varying the amount of fuel. The flow chart for solution phase sol-gel autocombustion method for synthesizing Ni-Zn ferrite nanoparticles and the post-synthesis characterizations carried out on the samples are shown in Figure 3 and typical images of sol-gel autocombustion process in different stages of synthesis are shown in Figure 4. As it has been mentioned in the previous section that the substituted Ni-Zn ferrite nanoparticles are very attractive materials for high frequency electronic device applications due to their high resistivity coupled with interesting magnetic properties.

Besides, this method is also suitable for preparation of all types of ferrites including cobalt based ferrites in different size scales simply by changing the metal salts to fuel ratio and also by changing the type of fuel, namely citric acid, glycine, tartaric acid, PVA, urea, etc., which forms basis for subsequent fabrication of self composites using the constituents of the same composition but in different particle sizes for magnetostrictive applications.

Though the as-synthesized ferrite nanoparticles are indicative of spinel crystal structures, the powders were further heat treated at 400 °C for 1hr to remove any surface adsorbed moisture or such impurities and also obtain better crystallization in the structure. It had been reported in the literature that the heat treated material exhibits good crystallinity and optimum magnetic properties.

Fig. 4. Typical images of a) nitrate-citrate solution, b) gel, c) burnt ash flakes after auto-combustion, and d) as-synthesized fluffy loose powder of ferrite nanoparticles.

2.2 Characterization techniques

Before making any further studies so as to test the material performance and application potential, it is often necessary to characterize the samples to ensure that the method of synthesis employed was the correct one and that the materials synthesized were exactly the ones intended to be prepared. For this purpose, the synthesized magnetic/ferrite nanoparticles of the basic compositions, whose properties were well established and widely reported in the literature, have been put to different structural and magnetic characterizations.

2.2.1 Structure characterization

X-ray diffraction patterns of the as-synthesized magnetite nanoparticles (S1 in the figure) confirm the formation of cubic inverse spinel structure, which is consistent with the standard JCPDS card no.00-019-0629 for magnetite, and the peaks can be indexed at the 2θ values of 30.1°, 35.4°, 37.0°, 43.0°, 53.39°, 56.9°, 62.6°, corresponding to the crystal planes of (220), (311), (222), (400), (422), (511), (440), respectively [25]. It is further reported that the X-ray diffraction data of the other two samples with silica coating shows, in addition to the above peaks for the core magnetite nanoparticles, a broad peak at $2\theta = 15\text{-}25°$ which can be ascribed to the amorphous silica. Furthermore, the relative intensity of the peaks, which reflect the magnetite phase, decreases in both the silica coated cases. Similar results of decreasing the intensity of magnetite phase peaks and the presence of broad peak for the amorphous silica phase for coating magnetite with silica have been reported [19].

Similarly, sol-gel autocombustion synthesized Ni-Zn ferrite nanoparticles too exhibit the formation of the spinel phase corresponding to the crystal planes of (220), (311), (222), (400), (422), (511) and (440) [25]. However, the strong peaks with minimal background noise and the observation of no peaks of any other phases in this case clearly indicate the formation of fully crystalline high purity ferrite with a cubic spinel structure. Besides, elemental analysis of the sample was also done using energy-dispersive X-ray spectroscopy (EDS). Typical EDS pattern taken on as-synthesized basic Ni-Zn ferrite nanoparticles is shown in Figure 5.

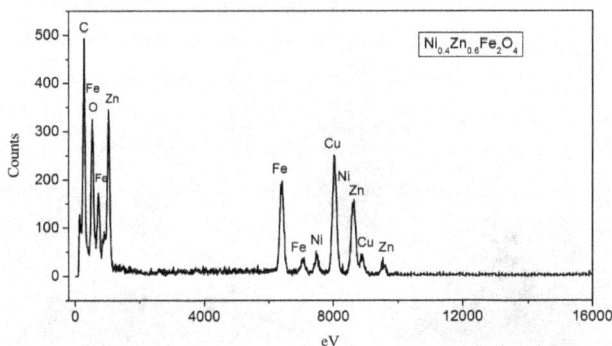

Fig. 5. Typical EDS pattern of synthesized basic Ni-Zn ferrite sample.

The EDS pattern quantitatively describes the presence of Ni, Zn, Fe and O elements in the respective sample used in the measurements. The peaks related to Cu and C are due to the source of incident Cu K_α radiation and carbon grid, respectively. Since the observed pattern contains no additional peaks for any of the elements other than those mixed for making the sample, it can be inferred that the sample adds no additional impurities during the synthesis. Further, though there exists some overlapping of the peaks related to different elements present in the sample, the analysis of the pattern approximately provided estimation of the elemental concentrations as per the specifications mixed at the time of preparation; thus confirming the compositions of the ferrite samples in correct proportions in the final products. Therefore, it is presumed from the figure that the targeted compositions are synthesized as per the chemical equations only which ensure that the method employed for synthesis is accurate.

2.2.2 Morphology characterization

Typical high resolution TEM images of the sonochemically synthesized Fe/Fe_3O_4 core shell nanoparticles and polyol synthesized Mn-Zn ferrite nanoparticles are shown in Figure 6(a,b). The dark part in the contrast image (a) represents Fe core and light part reflects the thickness of the shell. When the oxidation of the sample is more in air atmosphere, it results in increasing the thickness of the shell and a corresponding decrement in the core diameter. The same approach was used to obtain core-shell nanoparticles in other systems also including those of Fe_3O_4/SiO_2 and FeCo/ferrite core shell nanoparticles with the use of sonochemically synthesized Fe_3O_4 and $CoFe_2O_4$ nanoparticles in the first step. And, the morphological characteristics of $CoFe_2O_4$ nanoparticles whose images have been shown in Fig.2 earlier in three different preparatory conditions leads to wide variations in estimated average particle sizes depending on the aqueous mixing conditions.

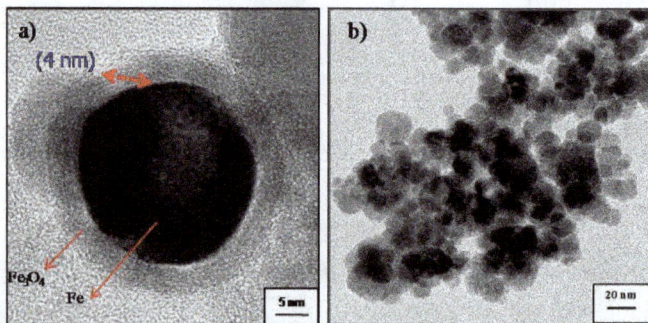

Fig. 6. TEM images of (a) Fe-Fe$_3$O$_4$ core-shell nanoparticles after oxidation in air by sonochemical method, and (b) Mn-Zn ferrite nanoparticles by polyol method (Courtesy: Abbas et al. [26]).

In another work of Fe_3O_4/SiO_2 particles [19], the mean particle size of the core magnetite nanoparticles before coating is nearly 10 nm, while the silica shell thickness after coating using Stober method has been increased to about 20 nm while resulting in nonuniform shell thickness. However, in order to improve the quality of coating, a modified Stober method was adopted which also results in non-uniform thickness but better surface coating. Polyethylene glycol was used in this approach and it could have played an important role in connecting the silica group with magnetite nanoparticles through formation of hydroxyl groups around magnetite surface and this in turn enhances the reaction with the silicon dioxide generated from TEOS to form Fe-O-Si bond. Additional experimental evidence in support of the formation of silica layer on the magnetite particle surface was also obtained in this work through the FTIR spectrum by observing peaks at around 542 cm^{-1} and 1130 cm^{-1}which are related to the vibration of Fe-O functional group, a characteristic peak of magnetite (Fe$_3$O$_4$) [19] and to the asymmetric vibration of Si-O-Si bond, respectively which confirmed the presence of silica in the coated samples.

Besides, the typical TEM images of polyol synthesized Mn-Zn ferrite nanoparticles are shown in Fig.6b. Compared to the size and morphology of the ferrite nanoparticles obtained by sonochemical method, the polyol method naturally produces smaller particles but in the form of aggregates with less dispersity, and the particles shapes and sizes are further dependent on synthesis conditions as well as the manganese concentration in the composition [27]. Furthermore, increasing the manganese content in the composition leads to enhancing the crystallinity of the particles with increase in the particle size. However, when the same composition is synthesized using sonochemical method, the nanoparticles are largely obtained in cube shapes. On the other hand, when these polyol and sonochemical approaches are prepared for synthesis of Ni-Zn ferrite nanoparticles, the sonochemical method was again effective in producing nanoparticles with very high crystallinity, narrow size distribution and uniform shapes [28].

2.2.3 Magnetic characterization

Magnetic strength of the synthesized nanoparticles is very important to find out their applicability because all these materials are primarily designed and made with an aim to make use of them in real applications. Interestingly, all the seed magnetite nanoparticles and Fe$_3$O$_4$/SiO$_2$ and Fe$_3$O$_4$/TiO$_2$ core-shell nanostructures exhibit strong magnetic properties as evident from their corresponding magnetic hysteresis loops [18,19]. Further, observed saturation magnetization values of the Fe/ Fe$_3$O$_4$and FeCo/ Fe$_3$O$_4$ core-shell nanostructures are 125 emu/g and 160 emu/g, respectively after 48 hours of oxidation. It implies that these samples are magnetically very strong to ensure that they are highly suitable for biomedical applicationsystems.

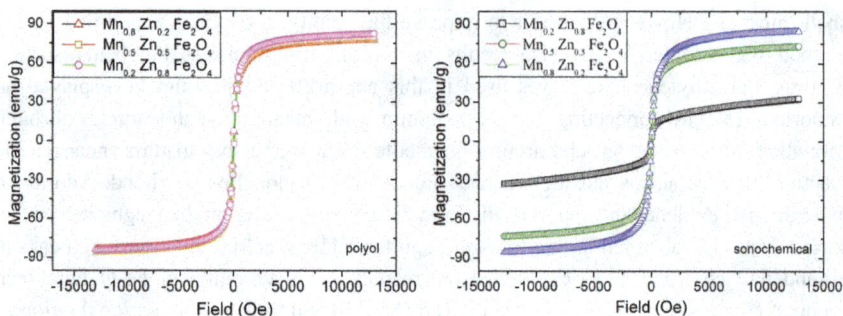

Fig. 7. Typical hysteresis loops of Mn-Zn ferrite nanoparticles synthesized by polyol and sonochemical methods (Courtesy: Abbas et al. [27]).

Similarly, magnetic properties of the ferrite nanoparticles synthesized by two different methods are shown as hysteresis curves in Figure 7. The saturation magnetization values for the polyol synthesized samples for which the mean particle sizes are in the range of 10-20 nm experience little variation with the composition. Their values remain more or less similar in the range from 77.5 emu/g to 82.3 emu/g when the Mn concentration is changed from 0.2 to 0.8. It can be inferred from the above that when the particle sizes are smaller, the corresponding magnetization variations are also smaller in spite of the change in Mn concentration in these Mn-Zn ferrite nanoparticles. On the other hand, the sonochemical synthesized nanoparticles experienced drastic change in saturation magnetization value with the change in Mn concentration because the corresponding mean particle sizes are about the size of 50 nm. Interestingly, all the samples have shown very little corecivities indicating superparamagnetic behavior.

3. High frequency device applications

Mn-Zn and Ni-Zn ferrites are the prime materials with widespread usage in large quantities for transformer cores, deflection yokes, inductors and antenna applications. The Mn-Zn ferrites, of course, with high permeability, high magnetization and low losses could serve as excellent core materials up to 1 MHz. The Ni-Zn ferrites, on the other hand, with moderate permeabilities and high resistivities expand their scope for applications as core materials even up to several MHz frequencies. For the miniaturized electronic device applications, since the throughput performance of the core is proportional to magnetic induction and frequency, the ferrite parameters that require utmost attention are high values of permeability, saturation magnetization and resistivity.

Improvements in these parameters would obviously help to increase the operating frequency of the electronic device beyond 1 MHz with acceptable core losses. Further, if the material is made in nano scales that to in single domain structures, domain wall resonance could be avoided which in turn enables the material for work even at high frequencies. Thus, researchers in this field studied Mn-Zn and Ni-Zn ferrites in various combinations by making systematic chemical modifications with an aim to improve their electrical and magnetic properties further.

In this regards, many reports were available in the literature devoting towards enhancing the performance of the Mn-Zn and Ni-Zn ferrites either by adopting optimum sintering schedules or by making suitable chemical modifications to each of these systems independently. Since the properties of ferrites strongly depend on the preparation methodology, heat treatment and the amount and type of impurities present in or added to them, rigorous efforts were also made to improve the performance by substituting/adding monovalent (Li, Ca, Na, K), divalent (Cu, Mg, Cd, Co), trivalent (In, Cr, Al, Sc, Bi), tetravalent (Si, Ti, Sn, Zr, Ge) and pentavalent (V, Nb, Ta) cations to the basic Mn-Zn/Ni-Zn ferrite composition. In the process of establishing the importance of synthesis conditions on the properties of ferrites, Koops [29] in the beginning studied a Ni-Zn ferrite composition using different sintering schedules and reported that the final products contain different levels of Fe^{2+} ion concentrations leading to wide variations in their resistivities and dielectric constants. Jain et al. [30] also reported that the variations in sintering conditions could lead to corresponding microstructural variations resulting in modifying the initial permeability as the grain size bears a linear relationship with the permeability [31].

A decrease in electrical resistivity of Ni-Zn ferrites at higher sintering temperature was also reported by attributing the same to increased grain size and decreased porosity [32]. Rao et al. [33] reported significant variations both in dc electrical resistivity and dielectric properties of a Ni-Zn ferrite sintered at different temperatures and times. They attributed the variations in properties largely to the corresponding presence of Fe^{2+} ions resulting from variations in sintering schedule adopted for each of the samples. Dasgupta et al. [34] observed that a proper sintering environment is necessary not only to form the Mn-Zn ferrites but also to obtain fine grain size and improved magnetic properties. They further noticed milling and annealing in air atmosphere could lead to restrict even the formation of Mn-Zn ferrites as they found the presence of only Zn-ferrite by their XRD analysis.

In tune with the recent developments in physical and chemical nano preparation routes, some groups have also made attempts to improve the Mn-Zn or Ni-Zn ferrite properties by employing different preparation routes for synthesizing the initial ferrite powders. Praveena et al. [35] developed nanocrystalline Mn-Zn ferrites by employing microwave-hydrothermal method and claimed that these ferrites could perform up to 1 MHz frequency range with minimum power loss due to improved densities and better microstructures. Dasgupta et al. [36] investigated the magnetic behaviour of Mn-Zn ferrite nanoparticles synthesized through mechanosynthesis route. The sizes of the as-synthesized particles after being subjected to Rietveld refinement was estimated to be about 6-8 nm, and that was further increased up to 14-18 nm after annealing in argon atmosphere at 700°C. They observed that the Mn-Zn ferrite nanoparticles could be switched from ferromagnetic to paramagnetic with the decrease in particle size and the Mn content. Jeyadevan et al. [37] conducted an extended X-ray absorption fine structure study to determine the ionic distribution in Mn-Zn ferrites prepared by co-precipitation method. They attributed the variations in electromagnetic properties of ferrites synthesized by different processing routes to the variations in particle size as well as to the cationic distribution on octahedral and tetrahedral sites. They further reported that the Fe^{3+}, Mn^{2+} and Zn^{2+} ions due to their possessing of zero CFSE, their site preferences could be influenced by cations with non-zero CFSE for a particular lattice site.

Among various wet chemical methods, the co-precipitation and sol-gel techniques are widely popular. However, co-precipitation is highly pH sensitive and sol-gel is sophisticated requiring expensive alkoxide precursors. On the other hand, there is another simple method which combines sol-gel and combustion processes to be named as sol-gel autocombustion [38]. This process has the advantage of producing nano-sized homogeneous powders by using inexpensive precursors.

Attempts were also made by several groups with much success in the direction of addition or substitution of either magnetic or non-magnetic ions to the basic mixed ferrite to improve the most desired property at the cost of the least desired. The addition of small amounts of monovalent Ca and Na in Ni-Zn ferrites was known to increase the electrical resistivity due to the segregation of the additives at the grain boundaries [39]. Similarly, the presence of small amounts of Sb_2O_3 and ZrO_2 in the solid solution of Ni-Zn ferrites enhanced the sinterability and improved the permeability and its temperature dependence.Zaspalis et al. [40] investigated Mn-Zn ferrites by incorporating additives like CaO, Nb_2O_5 and CoO to improve the permeability. The calcium and niobium ions are grain boundary dopants, and the cobalt ions are bulk dopants with adequate amounts of induced anisotropy and domain wall pinning contributed accordingly to marginally improve the permeability. The addition of small quantity of Al_2O_3 in Ni-Zn ferrites resulted in a reduction of both structural and magnetic properties [41]. However, the presence of Al^{3+} ions in between Fe^{2+} and Fe^{3+} ions could partially block the hopping of electron exchange between them, as a result there observed an increase in resistivity and decrease in dielectric loss leading to applicability of the materials at higher frequencies.

It has been reported that in the study of Ni-Zn ferrites with different kinds of substituents the Ni^{2+} ions are mostly substituted with divalent ions such as Cu^{2+}, Cd^{2+}, Co^{2+}, Ca^{2+}, Mg^{2+}, Mn^{2+}, Be^{2+}, etc., and the Zn^{2+} ions are substituted with only a few divalent ions like Cu^{2+}, Be^{2+}, etc. [42]. These substitutions, however, have been reported to influence the microstructure, intrinsic properties, atomic diffusivity and sintering kinetics of the basic Ni-Zn ferrite. In one such example, the substitution of Mn^{2+} ions in place of Ni^{2+} ions in the Ni-Zn ferrite has been reported to greatly affect the physical properties by increasing the lattice parameter and grain size [43]. As a result, the electromagnetic properties were improved with increase in magnetization, initial permeability and resistivity, and the materials become promising for high-frequency applications. However, if the Mn ions are substituted for Fe ions in Ni-Zn ferrite, the Mn enters the lattice in trivalent state and improves the square loop properties as well as the initial permeability marginally because of its positive magnetostrictive contribution [44]. And, if the Mn and Ni ions together are substituted for Fe, the Mn enters the lattice as Mn^{4+} and behaves the same manner as other non-magnetic tetravalent cations such as Ti, Ge, Sn, Si and Ge behave to degrade the magnetic properties of the basic Ni-Zn ferrite [45].

Co-precipitation method was used by Venkataraju et al. [46] to synthesize nanosized Mn-Zn ferrite. They substituted Ni for Mn in this study and observed a decrease in lattice constant and magnetization compared to the bulk values for the same ferrite. They explained their results in terms of spin canting and deviation in site preferences of the cations in the nano state from their normal bulk cationic distribution. Addition of V_2O_5 and Nb_2O_5 to the basic Ni-Zn and Mn-Zn ferrites had been reported as a corrective measure by several groups [47-49] to decrease the power losses and to raise the frequency at which the material can be operated safely in high frequency transformer applications. Effect of NiO doping on the properties of Mn-Zn ferrites was investigated by Sun et al. [50] and reported that the NiO helped to enhance the electrical resistivity, and thereby decreased the eddy-current loss. However, it caused the hysteresis loss to increase monotonically with the increased addition of NiO.

Arulmurugan et al. [51] investigated the influence of zinc substitution on nanocrystalline Mn-Zn and Co-Zn ferrite powders synthesized by co-precipitation method. They claimed that particle size and magnetic properties of these ferrites depend strongly on the processing route. The size of the nanoparticles has been observed to decrease with increasing Zn concentration in both the ferrite systems. Bueno et al. [52] synthesized Mn substituted Ni-Zn ferrites by employing nitrate-citrate precursor route. The manganese substitution in this ferrite system has been observed to increase lattice constant, saturation and remanence magnetic induction while decreasing the coercivity.

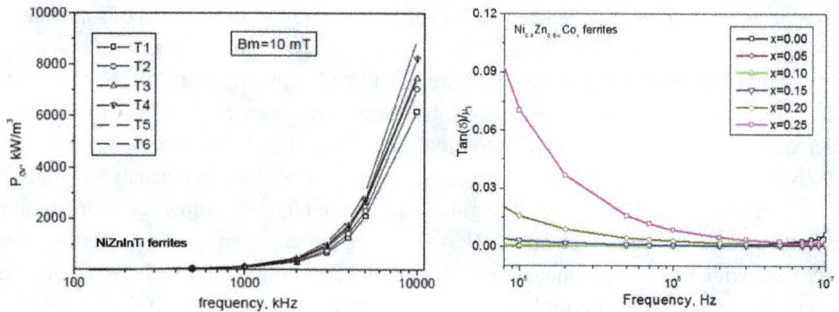

Fig. 8. Core losses versus frequency of conventionally prepared NiZnInTi ferrites and relative loss factor versus frequency of sol-gel autocombustion synthesized NiZnCo ferrites (Courtesy: Rao et al. [53,54]).

All these studies indicate that the properties of ferrites are strongly dependent upon the processing parameters and impurity content as well. Though it is difficult to prepare a ferrite with all good properties, it is possible to formulate the required mechanisms to improve one property at the cost of the other by suitably employing appropriate synthesis route, heat treatment and chemical modifications depending upon the application. Our recent studies, which attempts to utilize the advantages of both processing conditions and chemical substitutions in Ni-Zn ferrites, reported improved performance in core losses up to several megahertz [55].

Fig. 8 shows the curves of frequency dependent core losses (P_{cv}) and relative magnetic loss factors (tan δ/μ_i) for different concentrations of conventionally prepared Ni-Zn-In-Ti ferrites and sol-gel autocombustion synthesized Ni-Zn-Co ferrite nanoparticles, respectively. In the conventional samples, the core losses are minimal up to 2 MHz only whereas in the nano ferrite samples the relative loss factor remains low even up to 10 MHz indicating that the composition as well as the preparation route have profound influence on the frequency response of the material. Moreover, it suggests that though the Ti substitutions are able to contain the core losses upto a few MHz, the Co substitutions in Ni-Zn ferrites coupled with nano dimensions are more effective in reducing the losses, which could perhaps be due to increased uniformity, structural homogeneity and fine grained microstructures obtained from the advantages offered by the sol-gel autocombustion method in preparation of the initial powders. Furthermore, the most interesting aspect of this study is observation of low losses in Co substituted Ni-Zn ferrites in spite of large decrease in permeability compared to the undoped Ni-Zn ferrite

[55].This could perhaps be due to improvements in their corresponding anisotropy constant and resistivity parameters.

4. Magnetostrictive applications

Cobalt ferrites and their composites exhibit very interesting magnetostrictive properties and thus they find ample scope for use in various sensing and actuating applications. Early focus on the studies of magnetostrictive substituted cobalt ferrites was largely devoted to investigations on single crystals as there observed a linear increase in magnetostriction with cobalt content [56,57]. Nevertheless, studies on polycrystalline Co ferrites are relatively simple and less expensive from the technological point of view and thus several investigations were made on polycrystalline cobalt ferrites. Substitutions of Mn or Si in polycrystalline Co ferrites resulted in considerable decrease in magnetostriction and this has been attributed to the positive magnetostrictive contribution of manganese and magnetic dilution of B-sublattice [58,59]. Of course, some contrasting results were also observed in Co ferrite systems while reporting maximum values of strain derivative for lower Mn concentration by Jiles et al. [60] and for higher Mn/Si concentrations by Caltun et al. [61]. This discrepancy has been explained by Caltun et al. on the basis of decreasing anisotropic contribution of cobalt due to the cobalt migration to tetrahedral site with Mn/Si concentration. A complete reversal of sign of bulk magnetostriction has been observed in polycrystalline Co-Cr ferrites with increasing Cr substitution [62], which has been attributed to the aggregate contribution of magnetostriction in different crystallographic directions in polycrystalline cobalt ferrite.

In a series of studies, a gradual decrease in saturation magnetization has been reported in case of $CoMn_xFe_{2-x}O_4$, $CoCr_xFe_{2-x}O_4$ and $CoGa_xFe_{2-x}O_4$ systems by Jiles et al. [62,63] due to dilution of the strength of the magnetic ions. Whereas Caltun et al. [64] noticed a smaller decrease in magnetization up to $x = 0.2$ and then observed an increase up to $x = 0.4$ in $CoMn_xFe_{2-x}O_4$ system. This has been attributed to the preferred cationic distributions such as cobalt migration from octahedral to tetrahedral site in proportion to manganese content. Bhame and Joy[65,66] reported that magnetostriction and magnetization were increased with increasing manganese concentration up to $x = 0.3$ in $Co_{1-x}Mn_xFe_2O_4$ and reported a gradual decrease in magnetostriction and magnetization with increasing concentration of manganese in $CoMn_xFe_{2-x}O_4$ systems. Duong et al. [67] studied nano sized cobalt ferrite and cobalt zinc ferrite prepared by citrate gel method, with the particles in the size range of 40 nm, and reported that the longitudinal and transverse magnetostrictions are 130 and 70 ppm, respectively. The low values are attributed to disordered magnetic phase.

Recently, Mohaideen and Joy reported magnetostriction coefficient of 400 ppm and strain derivative of 2×10^{-9} m/A in cobalt ferrite self composites by mixing nano and micron sized powders in different ratios [68,69]. Wang et al. [70] studied magnetically oriented polycrystalline $CoFe_2O_4$ samples and reported large values of maximum magnetostriction and strain derivatives. Interesting aspect of their study was they obtained the results by annealing the samples well before sintering when they were in slurry form. Recently, Shyam et al. [71] explored spinel-perovskite dual phase cobalt ferrites to obtain significant changes in structural and magnetic properties which could obviously modify the magnetostrictive strength of the material. In our earlier study [72], highest strain derivative value 4.5×10^{-9} m/A and nearly 100 ppm magnetostriction coefficient value were observed in $Co_{1-x}Cu_xFe_2O_4$ system for $x = 0.15$ and the particle size is 50 nm for the sintered samples, the particle size is observed to increase with the copper content up to $x = 0.15$. The magnetostriction has been found to increase with the chromium content in $CoCr_xFe_{2-x}O_4$ from 160 to 230 ppm throughout the concentration.

It has been made clear from the above that the magnetostrictive properties of sintered products depend on initial particle size. The initial particle sizes of the samples are a deciding factor for improving the magnetostrictive properties. In addition, it has been shown in many works that the magnetic field annealing is effective for enhancing the magnetostriction coefficient as well as strain derivative of the cobalt ferrites [73,74] which could perhaps be due to parallel alignment of induced easy axis to that of annealing magnetic field and affecting magnetization processes as well as domain configuration. Mohaideen and Joy [68] reported enhanced magnetostriction and strain derivative in Mn substituted cobalt ferrites derived from nanocrystalline powders after magnetic field annealing. Mohaideen and Joy also reported further improvements in magnetostriction and strain derivative parameters when they investigated self-composites from nanosized and bulk powders with different particle sizes of the same material as components. It can be inferred from the above that the self-composites, i.e. by mixing the combinations of nano powders derived from combustion method using citric acid, glycine, etc., and micron sized powders from standard solid state method in different ratios, help to optimize the desired material performance by displaying higher magnetostriction coefficient and strain derivative when compared to the sintered products obtained from the individual powders and thus make themselves highly suitable for sensor and actuator applications.

5. Biomedical applications

Magnetic nanoparticles, particularly the magnetite (Fe_3O_4), are known to be very promising for biomedical applications due to their excellent physical properties, nontoxicity and high chemical stability in comparison to the metallic Fe and Co nanoparticles [20]. Small size (< 40 nm), uniform size distribution, superparamagnetic nature, solubility in water along with high magnetic moment are some of the physical parameters that are required to make these materials attractive for applications like targeted drug delivery, hyperthermia and magnetic resonance imaging enhancement. Magnetic nanoparticles are also used as magnetic labels for applications such as cell separation, manipulation and biomolecule detection in which the particle sizes can have the flexibility to be slightly larger to accommodate the presence of multiple ligands on the particle surface and to achieve multivalent interactions. In this case, the magnetic nanoparticles labels are encapsulated as biocompatible polymer beads with desired properties [75]. These labels/beads, as a result of coupling with adjacent encapsulated magnetic spins, tend to exhibit strong stray fields sufficient for efficient detection on successful bio-interaction. Nevertheless, the applicability of the magnetic nanoparticles in both these areas demand high saturation magnetization and biocompatibility as basic requirements, and thus the development of magnetite nanoparticles, which are known for their high biocompatibility and high magnetization, has become an obvious choice for intense investigations.

Polyol method which involves reduction of metal salts with a diol, typically ethylene glycol, diethylene glycol, or a mixture of both is believed to be one of the most appropriate methods for synthesis of hydrophilic nanoparticles [76]. The polyol process also helps to dispense the use of surfactant because poly ethylene glycol (PEG) is capable of playing a triple role as high-boiling solvent, reducing agent, and stabilizer to efficiently control the particle growth as well as inter-particle aggregation through steric interactions [77]. Thus, many research groups carried out works to synthesize monodisperse, water soluble magnetite nanoparticles with saturation magnetizations in the range from 45 to 77 emu/g using different polyols while keeping deoxygenated protection of the reaction [78,79]. Recently, Abbas et al. [20] reported a one-pot facile polyol method for the preparation of high magnetization hydrophilic Fe_3O_4 nanoparticles without using any surfactant and deoxygenated conditions.

In addition, magnetically soft metallic iron (Fe) and FeCo alloy nanoparticles (NPs) are also very attractive because they possess high magnetization values of 218 and 240 emu/g, respectively [25]. Though the superparamagnetic Fe nanoparticles and their stable dispersions with high magnetic moment are predicted to be useful in applications like bio-separation, bio-sensing, drug delivery and magnetic resonance imaging contrast enhancement, their scope is limited for reasons of toxicity and instability properties. However, iron nanoparticles compounds of Fe/ferrite and FeCo/ferrite in the form of core/shell nanoparticles are desirable since the metallic Fe and FeCo alloy cores produce the required high magnetic moment properties while the ferrite shell takes care of the stability as well as biocompatibility properties. Thus, in sight of the diverse applications, many groups have made efforts recently in synthesizing the Fe/ferrite and FeCo/ferrite core-shell nanoparticles using thermal decomposition and chemical reduction methods [25]. More recently, α-Fe/Fe$_3$O$_4$ nanocomposites with lamellar structures were also synthesized using hydrothermal method [80]. Though the above mentioned works succeeded in producing Fe/Fe$_3$O$_4$ nanostructures with good magnetic properties, some of these methods involve either complicated synthesis procedures or consume much amounts of surfactants. Keeping in mind the above, Abbas et al.[25] reported a facile approach for synthesizing high magnetic moment core/shell nanostructures of Fe/ferrite and FeCo/ferrite nanoparticles without the need for any surfactant or complicated process. They further reported that their materials are fully crystalline while the ferrite shell could well protect the metallic Fe and FeCo cores from deep oxidation and the metallic cores were stable in hexane and water dispersions, thus the obtained core-shell structures of Fe/ferrite FeCo/ferrite nanoparticles with desired stability are highly efficient for bio-separation, drug delivery and high sensitive bio-detection applications.

6. Conclusions

Magnetic nanoparticles and surface coated core-shell nanostructures have been the most sought after magnetic materials for they have a large scope for many applications including consumer electronics, automobile and biomedical fields. The technological importance and scientific necessity for exploring the nanomagnetic structures relevant to the above fields are briefly addressed in the first section. Simple, efficient and inexpensive methods for preparation and characterizations of the required materials have been briefly presented in the experimental details section. It follows from the above, an attempt was made to shed light in reporting the current status of research and application potential of magnetic nanoparticles and core-shell magnetic nanostructures in the fields of high frequency electronic devices, magnetostrictive sensors and actuators and biomedical applications. Emphasis was specially made on focussing compacted high resistive single domain ferrite nanoparticles for high frequency electronic devices, cobalt ferrite particle based self composites for magnetostrictive sensor and actuator applications, and core-shell nanoparticles for biomedical applications. It is intended that the contents of this work could well become a reference for the young researchers in these fields.

Acknowledgement

The authors wish to thank Dr. G.S.N. Rao for his kind help in the course of this work.

References

[1]   E.C. Snelling, Soft ferrites: Properties and Applications, 2nd ed., Butterworths Publishing, 1989.

[2]   H. Setyawan, F. Fajaroh, W. Widiyastuti, S. Winardi, I. Wuled Lenggoro, N. Mufti, One-step synthesis of silica-coated magnetite nanoparticles by electrooxidation of iron in sodium silicate solution, Journal of Nanoparticle Research, 14 (2012) 807–815. https://doi.org/10.1007/s11051-012-0807-7

[3]   L. Caruana, A.L. Costa, M.C. Cassani, E. Rampazzo, L. Peodi, N. Zacceroni, Tailored SiO_2-based coatings for dye doped superparamagnetic nanocomposites, Colloids and Surfaces A: Physicochemical and Engineering Aspects, 410 (2012) 111–118. https://doi.org/10.1016/j.colsurfa.2012.06.027

[4]   T.Y. Leung, C.Y.Chan, C.Hu, J.C.Hu, P.K.Wong, Photocatalytic disinfection of marine bacteria using fluorescent light, Water Research, 42 (2008) 4827-4837. https://doi.org/10.1016/j.watres.2008.08.031

[5]   X. Yu,S.Liu,J.Yu, Superparamagnetic γ-Fe_2O_3@SiO_2@TiO_2 composite microspheres with superior photocatalytic properties, Applied Catalysis B: Environmental, 104 (2011) 12–20. https://doi.org/10.1016/j.apcatb.2011.03.008

[6]   C. Hui, C. Shen, J. Tian, L. Bao, H. Ding, C. Li, Y. Tian, X. Shi, H.J. Gao, Core-shell Fe_3O_4@SiO_2 nanoparticles synthesized with well-dispersed hydrophilic Fe_3O_4 seeds, Nanoscale, 3 (2011) 701–705. https://doi.org/10.1039/C0NR00497A

[7]   M.A.Willard, L.K. Kurihara, E.E. Carpenter, S. Calvin, V.G. Harris, Chemically prepared magnetic nanoparticles, International Materials Reviews, 49 (2004) 125-170. https://doi.org/10.1179/095066004225021882

[8]   Q.A. Pankhurst, J. Connolly, S.K. Jones, J. Dobson, Applications of magnetic nanoparticles in biomedicine, Journal of Physics D: Applied Physics, 36 (2003) R167-R181. https://doi.org/10.1088/0022-3727/36/13/201

[9]   S.L. Tie, H.C. Lee, Y.S. Bae, M.B. Kim, K. Lee, C.H. Lee, Monodisperse Fe_3O_4/Fe@SiO_2 core/shell nanoparticles with enhanced magnetic property, Colloids and Surfaces A: Physicochemical and Engineering Aspects, 293 (2007) 278–285. https://doi.org/10.1016/j.colsurfa.2006.07.044

[10]   J.L.Snoek, New Development in Ferromagnetic Materials. Elsevier, New York, 1947.

[11]   J. Smit, H. P. J. Wijn, Ferrites, Philips Technical Library, Netherlands, 1959.

[12]   C.C.H. Lo, A.P. Ring, J.E. Snyder, D.C. Jiles, Improvement of magnetomechanical properties of cobalt ferrite by magnetic annealing, IEEE Transactions on Magnetics,41 (2005) 3676-3678. https://doi.org/10.1109/TMAG.2005.854790

[13]   S.H. Song, C.C.H. Lo, S.J. Lee, Magnetic and magnetoelastic properties of Ga-substituted cobalt ferrite, Journal of Applied Physics,101 (2007) 09c517.

[14]   O.F. Caltun, G.S.N. Rao, K.H. Rao, B. Parvatheeswara Rao, Cheol Gi Kim, Chong-Oh Kim, I. Dumitru, N. Lupu, H. Chiriac, High magnetostrictive cobalt ferrite for sensor application, Sensor Letters, 5 (2007) 1-3. https://doi.org/10.1166/sl.2007.027

[15]   O.F. Caltun, G.S.N. Rao, K.H. Rao, B. Parvatheeswara Rao, H.L.Wamocha, H. Hamdeh, Influence of silicon and cobalt substitutions on magnetostriction coefficient of cobalt ferrite, Hyperfine interactions, 184 (2008) 179-184. https://doi.org/10.1007/s10751-008-9786-6

[16]   S. Santra, R. Tapec, N. Theodoropoulou, J. Dobson, A. Hebard, W.H. Tan, Synthesis and Characterization of Silica-Coated Iron Oxide Nanoparticles in Microemulsion: The Effect of Nonionic Surfactants, Langmuir, 17 (2001) 2900–2906. https://doi.org/10.1021/la0008636

[17]   W. Stober, A. Fink, E. Bohn, Controlled growth of monodisperse silica spheres in the micron size range, Journal of Colloid and Interface Science, 26 (1968) 62-69. https://doi.org/10.1016/0021-9797(68)90272-5

[18]   Mohamed Abbas, B. Parvatheeswara Rao, Venu Reddy, CheolGi Kim, Fe_3O_4/TiO_2 core/shell nanocubes: Single-batch surfactantless synthesis, characterization and efficient catalysts for methylene blue degradation, Ceramics International, 40 (2014) 11177–11186. https://doi.org/10.1016/j.ceramint.2014.03.148

[19]   Mohamed Abbas, B. Parvatheeswara Rao, Md. Nazrul Islam, S.M. Naga, Migaku Takahashi, CheolGi Kim, Highly stable- silica encapsulating magnetite nanoparticles (Fe_3O_4/SiO_2) synthesized using single surfactantless- polyol process, Ceramics International, 40 (2014) 1379–1385. https://doi.org/10.1016/j.ceramint.2013.07.019

[20]   M. Abbas, B. Parvatheeswara Rao, S.M.Naga, Migaku Takahashi, CheolGi Kim, Synthesis of high magnetization hydrophilic magnetite (Fe_3O_4) nanoparticles in single reaction—Surfactantless polyol process, Ceramics International, 39 (2013) 7605-7611. https://doi.org/10.1016/j.ceramint.2013.03.015

[21] Wei-Chih Hsu, S.C. Chen, P.C. Kuo, C.T. Lie, W.S. Tsai, Preparation of NiCuZn ferrite nanoparticles from chemical co-precipitation method and the magnetic properties after sintering, Materials Science and Engineering: B, 111 (2004) 142–149. https://doi.org/10.1016/j.mseb.2004.04.009

[22] Anjali Verma, Ratnamala Chatterjee, Effect of zinc concentration on the structural, electrical and magnetic properties of mixed Mn–Zn and Ni–Zn ferrites synthesized by the citrate precursor technique, Journal of Magnetism and Magnetic Materials, 306 (2006) 313–320. https://doi.org/10.1016/j.jmmm.2006.03.033

[23] T. Slatineanu, A.R. Iordan, M.N. Palamaru, O.F. Caltun, V. Gafton, L. Leontie, Synthesis and characterization of nanocrystalline Zn ferrites substituted with Ni, Materials Research Bulletin, 46 (2011) 1455–1460. https://doi.org/10.1016/j.materresbull.2011.05.002

[24] K.C. Patil, Advanced ceramics: Combustion synthesis and properties, Bulletin of Materials Science, 16 (1993) 533-542. https://doi.org/10.1007/BF02757654

[25] Mohamed Abbas, Md.Nazrul Islam, B. Parvatheeswara Rao, K.E. AbouAitah, Cheol Gi Kim, Facile approach for synthesis of high moment Fe/ferrite and FeCo/ferrite core/shell nanostructures, Materials Letters,139 (2015)161–164. https://doi.org/10.1016/j.matlet.2014.10.078

[26] Mohamed Abbas, Sri Ramulu Torati, B. Parvatheeswara Rao, M.O. Abdel-Hamed, CheolGi Kim, Size controlled sonochemical synthesis of highly crystalline superparamagnetic Mn–Zn ferrite nanoparticles in aqueous medium, Journal of Alloys and Compounds,644 (2015) 774–782. https://doi.org/10.1016/j.jallcom.2015.05.101

[27] B. Parvatheeswara Rao, Chong-Oh Kim, CheolGi Kim, I. Dumitru, L. Spinu, O. F. Caltun, Structural and magnetic characterizations of coprecipitated Ni–Zn and Mn–Zn ferrite nanoparticles, IEEE Transactions on Magnetics, 42 (2006) 2858-2860. https://doi.org/10.1109/TMAG.2006.879901

[28] Mohamed Abbas, B. Parvatheeswara Rao, Cheol Gi Kim, Shape and size-controlled synthesis of Ni Zn ferrite nanoparticles by two different routes, Materials Chemistry and Physics, 147 (2014) 443-451. https://doi.org/10.1016/j.matchemphys.2014.05.013

[29] C.G. Koops, On the dispersion of resistivity and dielectric constant of some semiconductors at audio frequencies, Physical Review, 83 (1951) 121-125. https://doi.org/10.1103/PhysRev.83.121

[30] G.C. Jain, B.K. Das, N.C. Goel, Grain growth during sintering of manganese-zinc-iron ferrites, Indian Journal of Pure & Applied Physics,14 (1976) 87-92.

[31] A. Globus, P. Duplex, Separation of susceptibility mechanisms for ferrites of low anisotropy, IEEE Transactions on Magnetics,MAG-2 (1966) 441-445. https://doi.org/10.1109/TMAG.1966.1065867

[32] B. Parvatheeswara Rao, P.S.V. Subba Rao, A. Lakshman, K.H. Rao, Influence of sintering conditions on the microstructural and electrical properties of Ni-Zn ferrites, Journal of the Magnetics Society of Japan, 22 (1998) (S1) 83-85.

[33] B. Parvatheeswara Rao, K.H. Rao, Effect of sintering conditions on resistivity and dielectric properties of Ni-Zn ferrites, Journal of Materials Science, 32 (1997) 6049-6054. https://doi.org/10.1023/A:1018683615616

[34] S. Dasgupta, J. Das, J. Eckert, I. Manna, Influence of environment and grain size on magnetic properties of nanocrystalline Mn–Zn ferrite, Journal of Magnetism and Magnetic Materials, 306 (2006) 9-15. https://doi.org/10.1016/j.jmmm.2006.02.266

[35] K. Praveena, K.Sadhana, S.Bharadwaj, S.R.Murthy, Development of nanocrystalline Mn–Zn ferrites for high frequency transformer applications, Journal of Magnetism and Magnetic Materials, 321 (2009) 2433-2437. https://doi.org/10.1016/j.jmmm.2009.02.138

[36] S. Dasgupta, K.B. Kim, J. Ellrich, J. Eckert, I. Manna, Mechano-chemical synthesis and characterization of microstructure and magnetic properties of nanocrystalline $Mn_{1-x}Zn_xFe_2O_4$, Journal of Alloys and Compounds, 424 (2006) 13-20. https://doi.org/10.1016/j.jallcom.2005.12.078

[37] B. Jeyadevan, K. Tohji, K. Nakatsuka, A. Narayanasamy, Irregular distribution of metal ions in ferrites prepared by co-precipitation technique structure analysis of Mn–Zn ferrite using extended X-ray absorption fine structure, Journal of Magnetism and Magnetic Materials, 217 (2000) 99-105. https://doi.org/10.1016/S0304-8853(00)00108-6

[38] Z. Yue, Ji Zhou, L. Li, H. Zhang, Z. Gui, Synthesis of nanocrystalline NiCuZn ferrite powders by sol–gel auto-combustion method, Journal of Magnetism and Magnetic Materials, 208 (2000) 55-60. https://doi.org/10.1016/S0304-8853(99)00566-1

[39] N. Rezlescu, L. Rezlescu, P.D. Popa, E. Rezlescu, Influence of additives on the properties of a Ni–Zn ferrite with low Curie point, Journal of Magnetism and

Magnetic Materials, 215-216 (2000) 194-196. https://doi.org/10.1016/S0304-8853(00)00114-1

[40] V.T. Zaspalis, V. Tsakaloudi, M. Kolenbrander, The effect of dopants on the incremental permeability of MnZn-ferrites, Journal of Magnetism and Magnetic Materials, 313 (2007) 29-36. https://doi.org/10.1016/j.jmmm.2006.11.210

[41] H. L. Ge, Z. J. Peng, C. B. Wang, Z. Q. Fu, Effect of Al^{3+} doping on magnetic and dielectric properties of Ni–Zn ferrites by "one-step synthesis", International Journal of Modern Physics, B 25 (2011) 3881–3892. https://doi.org/10.1142/S0217979211101703

[42] Jozef Slama, A. Gruskova, M. Usakova, E. Usak, R. Dosoudil, Contribution to analysis of Cu-substituted NiZn ferrites, Journal of Magnetism and Magnetic Materials, 321 (2009) 3346-3351. https://doi.org/10.1016/j.jmmm.2009.06.024

[43] A.A. Sattar, H.M. El-Sayed, K.M. El-Shokrofy, M.M. El-Tabey, Effect of Manganese Substitution on the Magnetic Properties of Nickel-Zinc Ferrite, Journal of Materials Engineering and Performance, 14 (2005) 99-103. https://doi.org/10.1361/10599490522185

[44] H. Zhong and H. Zhang, Effects of different sintering temperature and Mn content on magnetic properties of NiZn ferrites, Journal of Magnetism and Magnetic Materials, 283 (2004) 247–250. https://doi.org/10.1016/j.jmmm.2004.05.029

[45] B. V. Bhise, M. B. Dongare, S. A. Patil, S. R. Sawant, X-ray infrared and magnetization studies on Mn substituted Ni-Zn ferrites, Journal of Materials Science Letters, 10 (1991) 922-924. https://doi.org/10.1007/BF00724783

[46] C. Venkataraju, G.Sathishkumar, K.Sivakumar, Effect of cation distribution on the structural and magnetic properties of nickel substituted nanosized Mn–Zn ferrites prepared by co-precipitation method, Journal of Magnetism and Magnetic Materials, 322 (2010) 230-233. https://doi.org/10.1016/j.jmmm.2009.08.043

[47] B. Parvatheeswara Rao, Cheol Gi Kim, Effect of Nb_2O_5 additions on the power loss of NiZn ferrites, Journal of Materials Science, 42 (2007) 8433–8437. https://doi.org/10.1007/s10853-007-1789-1

[48] B. Parvatheeswara Rao, Chong-Oh Kim, CheolGi Kim, Influence of V2O5 additions on the permeability and power loss characteristics of Ni–Zn ferrites, Materials Letters, 61 (2007) 1601–1604. https://doi.org/10.1016/j.matlet.2006.07.191

[49] S.H. Chen, S.C. Chang, C.Y. Tsay, K.S. Liu, I.N. Lin, Improvement on magnetic power loss of MnZn-ferrite materials by V_2O_5 and Nb_2O_5 co-doping, Journal of the European Ceramic Society, 21 (2001) 1931-1935. https://doi.org/10.1016/S0955-2219(01)00145-5

[50] K. Sun, Z. Lan, Z. Yu, L. Li, H. Ji, Z. Xu, Effects of NiO addition on the structural, microstructural and electromagnetic properties of manganese–zinc ferrite, Materials Chemistry and Physics, 113 (2009) 797-802. https://doi.org/10.1016/j.matchemphys.2008.08.052

[51] R. Arulmurugan, B. Jeyadevan, G. Vaidyanathan, S. Sendhilna, Effect of zinc substitution on Co–Zn and Mn–Zn ferrite nanoparticles prepared by co-precipitation, Journal of Magnetism and Magnetic Materials, 288 (2005) 470-477. https://doi.org/10.1016/j.jmmm.2004.09.138

[52] A.R. Bueno, M.L. Gregori, M.C.S. N´obrega, Effect of Mn substitution on the microstructure and magnetic properties of $Ni_{0.50x}Zn_{0.50x}Mn_{2x}Fe_2O_4$ ferrite prepared by the citrate–nitrate precursor method, Materials Chemistry and Physics, 105 (2007) 229–233. https://doi.org/10.1016/j.matchemphys.2007.04.047

[53] B. Parvatheeswara Rao, K.H. Rao, Distribution of In^{3+} ions in indium-substituted Ni–Zn–Ti ferrites, Journal of Magnetism and Magnetic Materials, 292 (2005) 44–48. https://doi.org/10.1016/j.jmmm.2004.10.093

[54] S. Ramesh, B.ChandraSekhar, P.S.V.SubbaRao, B.ParvatheeswaraRao, Microstructural and magnetic behavior of mixed Ni–Zn–Co and Ni–Zn–Mn ferrites, Ceramics International 40(2014)8729–8735. https://doi.org/10.1016/j.ceramint.2014.01.092

[55] S. Ramesh, Effect of Mn/Co Substitutions on the properties of nano and bulk Ni-Zn ferrites Ph.D. thesis, Andhra University, India, 2014.

[56] R.F. Pearson, The Magnetocrystalline Anisotropy of Cobalt-Substituted Manganese Ferrite, Proceedings of the Physical Society, 74 (1959) 505-512.

[57] R.D.Greenough, E.W.Lee, The magnetostriction of cobalt-manganese ferrite, Journal of Physics D: Applied Physics, 3(1970) 1595-1604. https://doi.org/10.1088/0022-3727/3/11/306

[58] J.C. Slonczewsky, Theory of Magnetostriction in Cobalt-Manganese Ferrite, Physical Review, 122 (1961) 1367-1372. https://doi.org/10.1103/PhysRev.122.1367

[59] G.S.N. Rao, B. Parvatheeswara Rao, O.F. Caltun, Cation Distribution of Cobalt-manganese Ferrite for Torque Sensor Applications, Materials Today: Proceedings, 2 (2015) 2491 – 2495. https://doi.org/10.1016/j.matpr.2015.07.192

[60] Y. Chen, J.E. Snyder, K.W. Dennis, R.W. McCallum, D.C. Jiles, Temperature dependence of the magnetomechanical effect in metal-bonded cobalt ferrite composites under torsional strain, Journal of Applied Physics, 87 (2000) 5798. https://doi.org/10.1063/1.372526

[61] G.S.N. Rao, O.F. Caltun, K.H Rao, P.S.V. Subba Rao, B. Parvatheeswara Rao, Improved magnetostrictive properties of Co–Mn ferrites for automobile torque sensor applications, Journal of Magnetism and Magnetic Materials, 341(2013) 60-64. https://doi.org/10.1016/j.jmmm.2013.04.039

[62] C.C.H. Lo, Compositional dependence of the magnetomechanical effect in substituted cobalt ferrite for magnetoelastic stress sensors, IEEE Transactions on Magnetics, 43 (2007)2367-2369. https://doi.org/10.1109/TMAG.2007.892536

[63] N. Wiriyal, A. Bootchanont, S. Maensiri, E. Swatsitang, X-ray absorption fine structure analysis of $Mn_{1-x}Co_xFe_2O_4$ nanoparticles prepared by hydrothermal method, Japanese Journal of Applied Physics, 53 (2014) 06JF09.

[64] O.F. Caltun, G.S.N. Rao, K.H. Rao, B. Parvatheeswara Rao, Ioan Dumitru, Chong-Oh Kim, Cheol Gi Kim, The influence of Mn doping level on magnetostriction coefficient of cobalt ferrite, Journal of Magnetism and Magnetic Materials, 316 (2007) e618-e620. https://doi.org/10.1016/j.jmmm.2007.03.045

[65] S.D. Bhame, P.A. Joy, Enhanced magnetostrictive properties of Mn substituted cobalt ferrite $Co_{1.2}Fe_{1.8}O_4$, Journal of Applied Physics, 99 (2006) 073901. https://doi.org/10.1063/1.2183356

[66] S.D. Bhame, P.A. Joy, Tuning of the magnetostrictive properties of $CoFe_2O_4$ by Mn substitution for Co, Journal of Applied Physics, 100 (2006) 113911. https://doi.org/10.1063/1.2401648

[67] G.V. Duong, R.S. Turtelli, N. Hanh, D.V. Linh, M. Reissner, H. Michor, J. Fidler, G. Wiesinger, R. Grössinger, Magnetic properties of nanocrystalline $Co_{1-x}Zn_xFe_2O_4$ prepared by forced hydrolysis method, Journal of Magnetism and Magnetic Materials, 307(2006) 313-317. https://doi.org/10.1016/j.jmmm.2006.03.072

[68]   K. Mohaideen, P.A. Joy, High magnetostriction and coupling coefficient for sintered cobalt ferrite derived from superparamagnetic nanoparticles, Applied Physics Letters, 101 (2012) 72405. https://doi.org/10.1063/1.4745922

[69]   K. Mohaideen, P.A. Joy, Enhancement in the Magnetostriction of Sintered Cobalt Ferrite by Making Self-Composites from Nanocrystalline and Bulk Powders, Applied Materials and Interfaces, 4 (2012) 6421-6425. https://doi.org/10.1021/am302053q

[70]   J. Wang, X. Gao, C. Yuan, J. Li, X. Bao, Magnetostriction properties of oriented polycrystalline $CoFe_2O_4$, Journal of Magnetism and Magnetic Materials,401(2016) 662–666. https://doi.org/10.1016/j.jmmm.2015.10.073

[71]   Shyam K. Gore, Santosh S. Jadhav, Vijaykumar V. Jadhav, S. M. Patange, Mu. Naushad, Rajaram S. Mane, Kwang Ho Kim, The structural and magnetic properties of dual phase cobalt ferrite, Scientific Reports, 7 (2017) 2524. https://doi.org/10.1038/s41598-017-02784-z

[72]   B. Chandra Sekhar , G.S.N. Rao,.O.F. Caltun, B. Dhanalakshmi, B. Parvatheeswara Rao, P.S.V. Subba Rao, Magnetic and magnetostrictive properties of Cu substituted Co-ferrites, Journal of Magnetism and Magnetic Materials,398 (2016) 59-63. https://doi.org/10.1016/j.jmmm.2015.09.028

[73]   R. W. McCallum, K. W. Dennis, D. C.Jiles, J. E. Snyder, Y. H. Chen, Composite magnetostrictive materials for advanced automotive magnetomechanical sensors, Low Temperature Physics, 27 (2001) 266-274. https://doi.org/10.1063/1.1365598

[74]   Y.X. Zheng, Q.Q. Cao, C.L. Zhang, H.C. Xuan, L.Y. Wang, D.H. Wang, Y.W. Du, Study of uniaxial magnetism and enhanced magnetostriction in magnetic-annealed polycrystalline $CoFe_2O_4$, Journal of Applied Physics,110 (2011) 043908. https://doi.org/10.1063/1.3624661

[75]   Q.S. Tang, D.S. Zhang, X.M. Cong, M.l. Wan, L.Q. Jin, Using thermal energy produced by irradiation of Mn–Zn ferrite magnetic nanoparticles (MZF-NPs) for heat-inducible gene expression, Biomaterials, 29 (2008) 2673-2679. https://doi.org/10.1016/j.biomaterials.2008.01.038

[76]   Mohamed Abbas, Md. NazrulIslam, B. Parvatheeswara Rao, T. Ogawa, Migaku Takahashi, Cheol Gi Kim, One-pot synthesis of high magnetization air-stable FeCo nanoparticles by modified polyol method, Materials Letters, 91(2013) 326–329. https://doi.org/10.1016/j.matlet.2012.10.019

[77]   F. Fievet, J.P. Lagier, M. Figlarz, Preparing Monodisperse Metal Powders in
       Micrometer and Submicrometer Sizes by the Polyol Process, Materials Research
       Society Bulletin, 14 (1989) 29-34.

[78]   F. Dang, N. Enomoto, J. Hojo, K. Enpuku, Sonochemical synthesis of
       monodispersed magnetite nanoparticles by using an ethanol–water mixed solvent,
       Ultrasonics Sonochemistry, 16 (2009) 649-654.
       https://doi.org/10.1016/j.ultsonch.2008.11.003

[79]   J.H. Bang, K.S. Suslick, Applications of ultrasound to the synthesis of
       nanostructured materials, Advanced Materials, 22 (2010) 1039-1059.
       https://doi.org/10.1002/adma.200904093

[80]   H. Wang, D. Udukala, T. Samarakoon, M. Basel, M. Kalita, G. Abayaweera,
       Nanoplatforms for highly sensitive fluorescence detection of cancer-related
       proteases, Photochemical and Photobiological Sciences, 13 (2014) 231–240.
       https://doi.org/10.1039/c3pp50260k

Chapter 8

Structural, Magnetic and Dielectric Properties of Aluminum Cobalt Substituted M−type Strontium Hexaferrites

Chetna C. Chauhan[1,a], Rajshree B. Jotania[2,b], Charanjeet Singh Sandhu[3,c]

[1]Institute of Technology, Nirma University, Ahmedabad – 382 481, Gujarat, India

[2]Department of Physics, University School of Sciences, Gujarat University, Ahmedabad – 380 009, Gujarat, India

[3]Department of Electronics and Communications, Lovely Professional University, Jalandhar–411 001, Punjab, India

[a]chetumakwana@gmail.com,[b]rbjotania@gmail.com,[c]rcharanjeet@gmail.com

Abstract

$SrCo_xAl_xFe_{(12-2x)}O_{19}$ (x = 0.0, 0.2, 0.4, 0.6, 0.8, 1.0) hexaferrites have been synthesized using a simple heat treatment method and characterized using various instrumental techniques such as FTIR, XRD, SEM, VSM and dielectric measurements. The XRD analysis reveals the formation of mixed phases of M-type hexaferrite and α-Fe_2O_3. The crystallite size is found in the range of 24-46 nm. The micrographs of typical samples show porous and agglomerated grains. The values of M_s, M_r, H_c decreased with the increase of Co–Al content. The values of the dielectric constant, tangent loss, AC conductivity and dielectric modulus were studied as a function of frequency.

Keywords

Strontium Hexaferrite, XRD, Saturation Magnetization, Coercivity, Dielectric Loss Tangent

Contents

1. Introduction

M−type strontium hexaferrite possesses a magnetoplumbite structure and is the prominent magnetic material due to its distinct magnetic properties such as high Curie temperature, a large value of magnetocrystalline anisotropy constant, high permeability, excellent chemical stability and corrosion resistance [1–4]. The crystal structure of M-type hexagonal ferrite can be considered as a superposition of spinel and hexagonal layers of $Fe_6O_8^{+2}$ and $MFe_6O_{11}^{-2}$, it has 24 Fe^{3+} ions per unit cell, that are distributed among five different interstitial sites: three octahedral sites 12k, 2a and $4f_2$, one tetrahedral site $4f_1$ and one trigonalbipyramidal site 2b. The 12k, 2a and 2b sites have up spins; while $4f_1$ and $4f_2$ sites have down spins because of the ferromagnetic coupling occurring due to the decrease in the distance of Fe–O–Fe and the angle between Fe–O–Fe approaching towards $180°$ [5]. The Fe ions at the 2b site provide the largest positive contribution to magnetocrystalline anisotropy, while those at the $4f_1$, $4f_2$, 2a sites provide medium to weak positive contributions, and a negative contribution is provided by Fe ions at the 12k site [6]. These magnetic materials find a vital role for the applications in microwave devices and electromagnetic wave absorber [7-8].

Strontium hexaferrite can be prepared by a variety of methods [9] like coprecipitation [10], microemulsion [11], conventional ceramic [12], and sol-gel [13]. The properties of hexagonal ferrites can be tuned by substituting cations during the synthesis. Hexagonal ferrites substituted with an appropriate proportion of ions can change structural, magnetic and dielectric properties of materials. The magnetic properties of strontium ferrite can be improved by substituting ferric ions with other suitable ions, such as Pb^{+2} [14], Cr^{+3} [15], Cu^{+2} [16] Al^{+3} [17-18] etc. It was reported that low level Co^{+2}–Al^{+3} substitution led to a greater control over magnetic parameters (M_s, M_r, H_c) in $SrFe_{12-x-y}Al_xCo_yO_{19}$ ($x = 1.0$ - 4.0, and $y = 0.0$ -2.0) series [19]. Jasbir Singh et al. [20] have studied tunable microwave absorptions in Co-Al substituted Ba-Sr hexaferrites and concluded that synthesized materials have potential as microwave absorbers. The Coercivity value has been enhanced by Al- substitution for Fe ions compared to pure M-type strontium hexaferrite [21].

In the present chapter, we report the effect of Co^{+2}–Al^{+3} substitution on structural, morphological, magnetic and dielectric properties of $SrCo_xAl_xFe_{(12-2x)}O_{19}$ ($0.0 \leq x \leq 1.0$) hexaferrites, prepared using a simple heat treatment technique.

2.   Experimental procedure

The simple heat treatment method was used to prepare $SrCo_xAl_xFe_{(12-2x)}O_{19}$ ($0.0 \leq x \leq 1.0$) hexaferrite samples. Stoichiometric proportion of strontium nitrate–$Sr(NO_3)_2 \cdot 6H_2O$ (≥ 99.0 % pure, Sigma–Aldrich), iron (III) nitrate –$Fe(NO_3)_3 \cdot 9H_2O$ (≥ 99.0 % pure, Sigma–Aldrich), aluminium   nitrate nonahydrate– $Al(NO_3)_3 \cdot 9H_2O$ (99.99 % pure, Sigma–Aldrich) and cobalt (II) nitrate hexahydrate–$Co(NO_3)_2 \cdot 6H_2O$ (ACS reagent, ≥ 98 % pure, Sigma–Aldrich) (all AR grade) were dissolved in deionized water separately. An aqueous solution of Polyvinylpyrrolidone–$(C_6H_9NO)_n$ (PVP 40, Sigma–Aldrich) was prepared separately in 100 ml of deionized water. As prepared solution of strontium nitrate, cobalt nitrate, aluminium nitrate and iron (III) nitrate was added one by one into the PVP 40 solution and stirred continuously for 2 h on a magnetic stirrer. The pH of the solution obtained was kept at 2. The mixed solution was heated to 80° C in order to evaporate the water content. The resulting dark orange brown solid was crushed for 15 min and heated at 650° C for 3 h to obtain the final product.

3.   Results and discussion

3.1   FTIR analysis

FTIR spectra of all samples were recorded at room temperature using an FTIR spectrometer (Bruker Tensor 27 Model). Fig. 1 represents the FTIR spectra of $SrCo_xAl_xFe_{(12-2x)}O_{19}$ ($0.0 \leq x \leq 1.0$) hexaferrites in the wave number range between 4000 cm^{-1} to 400 cm^{-1}. The FTIR spectra for all samples show two absorption bands v_1 and v_2 between 610 to 450 cm^{-1} [22], due to metal–oxygen (M–O) stretching vibrations among the tetrahedral and octahedral interstitial sites. The absorption band ~ 3400 cm^{-1} is observed in the $x = 1.0$ sample, showing the presence of a (O-H) group [23].

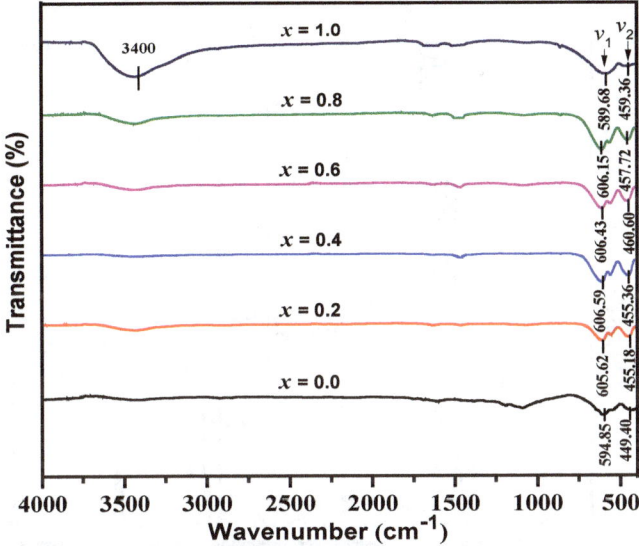

Fig. 1. FTIR spectra of SrCo$_x$Al$_x$Fe$_{(12-2x)}$O$_{19}$ (0.0 ≤ x ≤ 1.0) hexaferrites, heated at 650° C for 3 h.

3.2   Phase identification

X-ray diffraction patterns of heated samples were recorded on a Bruker D Z Phaser diffractometer (PW 1830) using Cu-K$_\alpha$ radiation (λ =1.5405 Å) with a scan rate of 2°/min and are shown in Fig. 2. The obtained XRD peaks were identified using 'Powder X' software. The diffraction peaks corresponding to the planes (006), (008), (110), (114), (108), (203), (205), (206), (301), (20 11), (221), (20 14), (317) matched with PDF# 841531 (a = 5.884 Å, c = 23.049 Å) and confirm formation of a hexagonal crystal structure, while (104) and (024) match with CAS registry no.1309–37–1, showing the presence of α-Fe$_2$O$_3$. For x = 0.0 to x = 0.8, the most intense peak is (114), whereas for x = 1.0, the most intense peak is (108). The most intense peak (114) is found to shift right with the substitution of Co–Al content (x) in SrCo$_x$Al$_x$Fe$_{12-2x}$O$_{19}$ as compared to the pure strontium hexaferrite (x = 0.0). It is interesting to note that the relative intensity of the (114) peak changes and this peak disappeared in the x =1.0 sample. Lattice parameters, unit cell volume, crystallite size, Bragg angle (2θ) of the most intense peak are listed in Table 1.

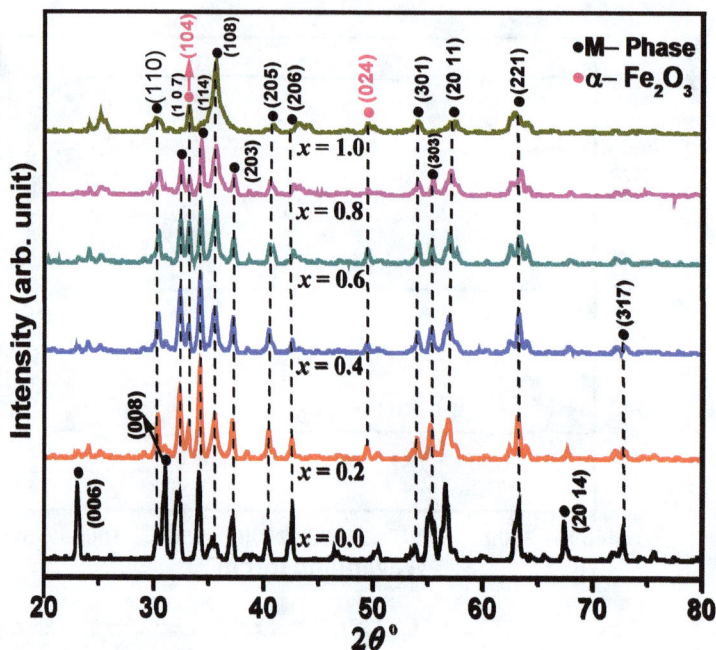

Fig. 2. X-ray diffraction patterns of SrCo$_x$Al$_x$Fe$_{(12-2x)}$O$_{19}$ (0.0 ≤ x ≤1.0) hexaferrites heated at 650° C for 3 h.

It is clear from Fig. 3 and Table 1 that the lattice constants, unit cell volume, and c/a ratio did not change appreciably in the range $0.0 \leq x \leq 0.8$, where as 'a' varied between 5.883 Å and 5.870 Å, 'c' changed from 23.037Å to 23.049 Å, while the unit cell volume decreased from 690.824 Å^3 to 687.506 Å^3 and the c/a ratio was found in between 3.925 and 3.947. The lattice parameters for the sample with x =1.0, however, it showed an appreciable changes with $^c/_a$ = 4.012; the obtained c/a ratio is lower for the x = 0.0 to 0.8 samples but it is higher for the x = 1.0 sample, than reported standard value (3.98) of M-type hexagonal structure [24]. This may be due to the difference in ionic radii of Co^{2+} (0.72 Å), Al^{3+} (0.51Å), and Fe^{3+} (0.64Å). The average crystallite size (D_{XRD}) of all samples was calculated using Debye-Scherrer formula [25] by considering the most intense Bragg's peak of XRD.

$$D_{XRD} = 0.9 \frac{\lambda}{\beta \cos \theta} \tag{1}$$

Where λ is the wavelength of Cu-K$_\alpha$ radiation, θ is the Bragg's angle of diffraction and β is the difference between FWHM and the instrumental broadening. The crystallite size varied from 22.89 to 45.83 nm.

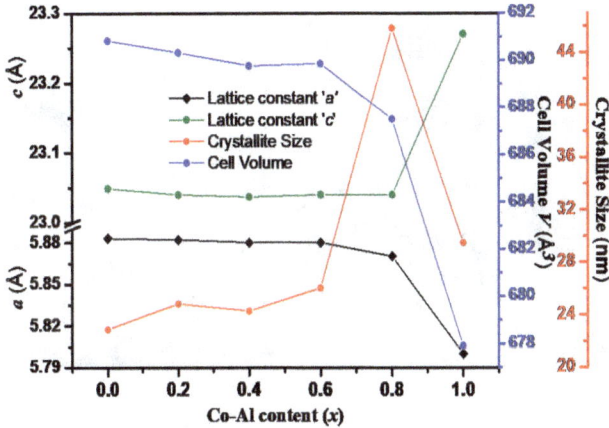

Fig. 3. Variation in lattice parameters, unit cell volume and crystallite size with Co–Al content (x) for $SrCo_xAl_xFe_{(12-2x)}O_{19}$ (0.0 $\leq x \leq$ 1.0) hexaferrites heated at 650° C for 3 h.

Table 1. Lattice parameters (a, c), unit cell volume (V), ratio (c/a), peak position of most intense peak and crystallite size (D_{XRD}) of $SrAl_xCo_xFe_{12-2x}O_{19}$ (0.0 $\leq x \leq$ 1.0) hexaferrites heated at 650° C for 3 h.

Co-Al content (x)	a (Å)	c (Å)	c/a	V (Å)³	Most Intense Peak Position (degree)	Most Intense Peak (hkl)	Crystallite size D_{XRD} (nm)
0.0	5.883	23.049	3.945	690.824	34.120	114	22.89±1.14
0.2	5.882	23.040	3.939	690.320	34.183	114	24.83±1.24
0.4	5.880	23.037	3.947	689.760	34.190	114	24.20±1.21
0.6	5.880	23.040	3.945	689.851	34.250	114	26.03±1.30
0.8	5.870	23.040	3.925	687.506	34.300	114	45.83±2.29
1.0	5.800	23.270	4.012	677.907	35.599	108	29.48±1.47

3.3 Morphological studies

Fig. 4 represents SEM micrographs of $x = 0.4$ and $x = 0.8$ samples. It is observed from the micrographs that both samples possess porous and agglomerated microstructure. Composition $x = 0.8$ ascribes more agglomeration and less porosity in comparison to $x = 0.4$. This agglomeration can be due to the magnetic dipole-dipole interactions [26-27].

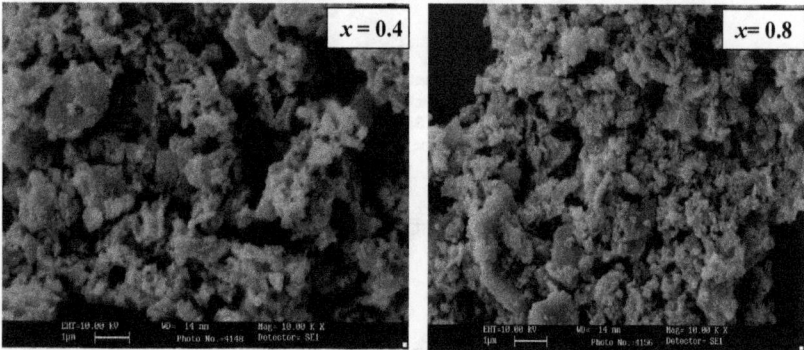

Fig. 4. SEM images of SrCo$_x$Al$_x$Fe$_{12-2x}$O$_{19}$ (x = 0.4 and x = 0.8) hexaferrites heated at 650° C for 3 h.

3.4 Magnetic properties

Fig. 5 depicts the M-H loops of SrCo$_x$Al$_x$Fe$_{12-2x}$O$_{19}$ ($0.0 \leq x \leq 1.0$) samples recorded at room temperature using a vibrating sample magnetometer (VSM, EG & G Princeton Applied Research instrument, Model–4500) under an applied field of up to 15 kOe. The Magnetic parameters are listed in Table 3. Variation of saturation magnetization (M_s), remanence magnetization (M_r) and coercivity (H_c) with Co– Al content (x) is shown in Fig.6.

It is clear from Fig. 6 and Table 3 that there is a decrease in M_r, M_s and H_c with Co–Al content (x). M_s varied from 43 to 24 emu/g, M_r changed from 24 to 7 emu/g and H_c decreased from 6546 to 801 Oe.

The magnetic behavior of M–type ferrites is controlled by the distribution of iron ions in the crystal lattice sites.

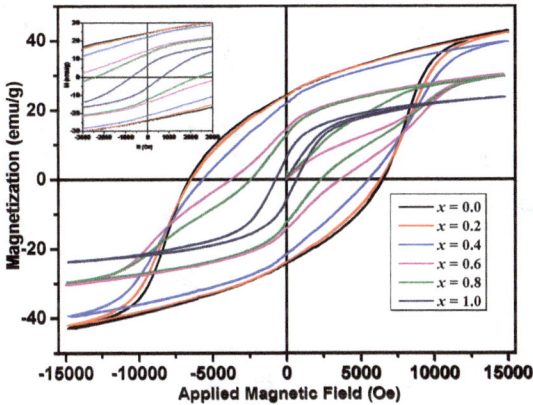

Fig. 5. Hysteresis loops of $SrCo_xAl_xFe_{12-2x}O_{19}$ $(0.0 \leq x \leq 1.0)$ hexaferritess heated at $650°C$ for 3 h.

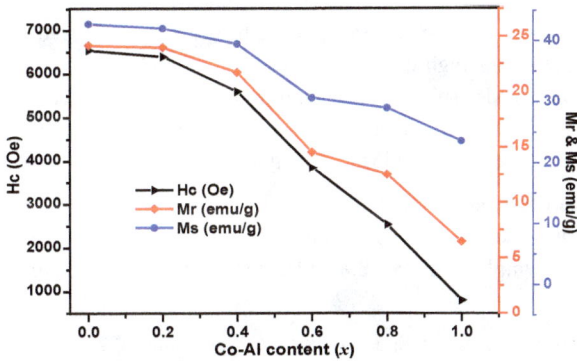

Fig.6. Variation of coercivity (H_c), saturation magnetization (M_s) and remanence magnetization (M_r) with Co-Al content (x) of $SrCo_xAl_xFe_{12-2x}O_{19}$ $(0.0 \leq x \leq 1.0)$ samples heated at $650°C$ for 3 h.

The net magnetization in M– ferrite can be expressed as:

$$M_S = M(12k + 2a + 2b)\uparrow - M(4f_1 + 4f_2)\downarrow \qquad (2)$$

223

It is clear from equation (2) that substitution of Fe^{3+} in spin up sites enhances net magnetization but in spin down sites reduces net magnetization. It is reported [28, 29] that non-magnetic Al^{3+} ions prefer to occupy 12k ($\uparrow$) sites and hence as results decrease in saturation magnetization. The difference in the magnetic moments of iron ions and the substituted $Al^{+3}-Co^{+2}$ ion may have diluted the magnetic interactions and resulted in the decrease of net magnetization.

The decrease in the coercive force can be attributed to the replacement of Fe^{+3} ions at $4f_2$ sites. The decrease in coercivity is also observed in Ba–Sr–Co–Zr ferrite [30].

It can be predicted that the increase in Co–Al substitution decreases the anisotropy field which in turn leads to the decrease in coercivity. The values of coercivity are greater than ($H_c > M_r/2$) half of the value of remanence magnetization, which makes the material a suitable candidate for high frequency applications [31,32].

The magnetic behavior hexaferrites depend on many factors like ionic radii and amount of substituents as well as the strength of magnetic interactions among cations. The formation of lattice defects due to substitution, random orientation of spins and the weaker super exchange mechanism seems to decrease the values of M_s, M_r and H_c [33]. The samples with $x < 0.4$ are characterized by high coercivity compared to typical values of ~ 4 kOe for SrM. This high coercivity, with moderate remanence of > 23.96 emu/g makes these compounds suitable for permanent magnet applications. At higher concentrations ($x > 0.8$) the coercivity drops to below 2.5 kOe which makes the compounds of potential applications for magnetic recording.

The value of squareness ratio (S) is found to be in the range of 0.47 to 0.57 for $x = 0.0$ to 0.6 samples, which depicts the formation of single domain isolated ferromagnetic particles [31]. The samples with $x = 0.8$ and 1.0 show $S < 0.5$, indicating multi-domain nature.

The Bohr magneton number for $SrCo_xAl_xFe_{12-2x}O_{19}$ ($x = 0.0, 0.2, 0.4, 0.6, 0.8$ and 1.0) hexaferrites was calculated using the formula:

$$n_B = \frac{M \times M_s}{5585} \tag{3}$$

Where M is the molecular weight of the sample and M_s is the saturation magnetization. It is clear from equation (3) that the number of Bohr magneton widely depends on saturation magnetization. It is evident from Table 3 that the values of Bohr magneton decrease with the increase in Co-Al content (x).

Table 3. Room temperature magnetic parameters of $SrCo_xAl_xFe_{12-2x}O_{19}$ (0.0 ≤ x ≤ 1.0) samples heated at 650° C for 3 h (Coercivity–H_c, Saturation Magnetization–M_s, Remanent Magnetization– M_r, squareness ratio– S, and Bohr magnetron–n_B measured at 15 kOe).

Co-Al content (x)	M_s (emu/g)	M_r (emu/g)	H_c (Oe)	$S = M_r/M_s$	n_B (μ_B)
0.0	42.845	24.163	6546.3	0.564	2.68
0.2	42.085	23.961	6397.9	0.570	2.56
0.4	39.519	21.716	5597.3	0.550	2.34
0.6	30.677	14.524	3486.7	0.473	1.76
0.8	29.055	12.524	2537.9	0.432	1.62
1.0	23.664	6.459	801.36	0.273	1.28

3.4 Frequency dependent dielectric properties

Dielectric characterization is an important tool to understand the electrical properties of materials. To study the dielectric characteristic the powder was pressed and pellets were prepared. The pellets were kept in between rectangular conducting plates to create an assembly similar to parallel plate capacitor. The dielectric measurements were recorded in the frequency range of 20 Hz to 2 MHz at room temperatureusing a precision LCR meter (Agilent E4980A).

The real part of dielectric signifies the quantity of energy stored in the dielectric material when kept under an AC electric field. The real part of permittivity (ε') of all samples have been calculated by using the formula:

$$\varepsilon' = \frac{Ct}{\varepsilon_o A} \tag{4}$$

Where t is the thickness of the pellet, A is area and ε_o is the permittivity in free space.

The variation of real part of dielectric constant with frequency is represented in Fig. 7. All the samples exhibit strong dielectric dispersion in the lower frequency region and the behavior is frequency independent in the higher frequency region. These higher values in the lower frequency region are due to the presence of grain boundary defects, voids, dislocations etc. [34]. The reduction in dielectric constant at lower frequency region can be explained on the basis of Koop's model [35]. According to this model, polycrystalline ferrites are assumed to be made up of conducting grains separated by non conducting grain boundaries. The accumulation of charge carriers near the grain boundaries results in the decrease in the polarization which in turn decreases the dielectric constant near the higher frequency region. Variation of dielectric constant (real- ε') with Co-Al content (x) is shown in Fig. 8.

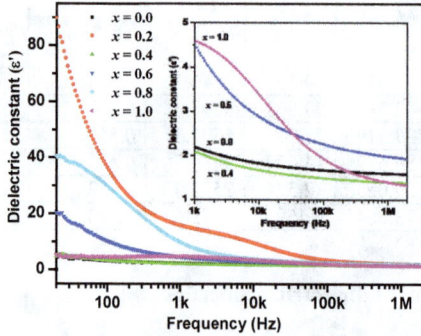

Fig. 7. Frequency dependent dielectric constant (real– ε) of $SrCo_xAl_xFe_{12-2x}O_{19}$ (0.0 $\leq$ x $\leq$ 1.0) samples heated at 650° C for 3 h.

The dielectric loss tangent (*tan δ*) of all samples was calculated by using relation:

$$tan\ \delta = \frac{\varepsilon''}{\varepsilon'}$$ (5)

Where ε' and ε'' are real and imaginary part of dielectric constant.

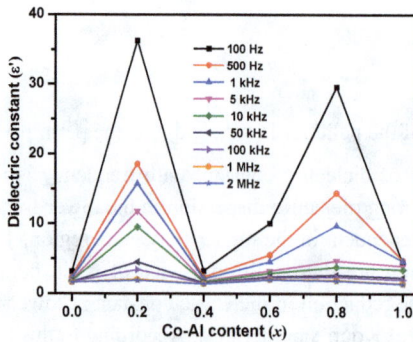

Fig.8. Variation of dielectric constant (ε') with Co-Al content (x) of $SrCo_xAl_xFe_{12-2x}O_{19}$ (0.0 $\leq$ x $\leq$ 1.0) samples heated at 650° C for 3 h.

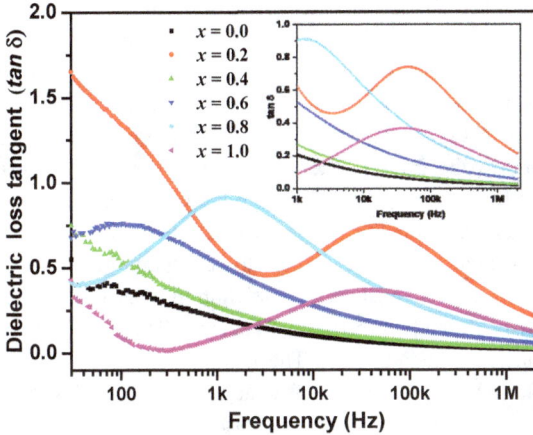

Fig.9. Frequency dependent dielectric loss tangent (tan δ) of $SrCo_xAl_xFe_{12-2x}O_{19}$ (0.0 ≤ x ≤ 1.0) hexaferrite samples heated at 650° C for 3 h.

The dielectric loss factor as a function of frequency in the range of 20 Hz to 2 MHz at room temperature is shown in Fig. 9. It is observed that $x = 0.2$, 0.8 and 1.0 samples show peaking or dielectric relaxation; while in other samples, dielectric loss tangent decreases with the increase in frequency. These phenomena can be explained by Koop's theory. Owing to the larger values of resistance near the grain boundary at low frequencies, electrons hopping from one site to another require higher energy values. This increases the loss in the low frequency region. In the high frequency region resistance is low. Hence a low energy is required for the electron to move from one site to another; so the energy loss is less [36, 37]. For $x = 0.2$, 0.8 and 1.0 samples, the loss tangent increases and then it decreases, this broadening in the curves with the increase in frequency is due to the orientation of the dipoles with the alternating field as reported earlier [37, 38]. The presence of broad relaxation peaks in the dielectric loss tangent curves is indicative of multiple relaxations.

The AC conductivity was calculated using the formula;

$$\sigma_{ac} = \omega\varepsilon_o\varepsilon'(\tan \delta) \tag{6}$$

Where $\omega = 2\pi f$ is the angular frequency and ε_0 is the permittivity of the free space.

Fig. 10 shows a variation of AC conductivity (σ_{ac}) with frequency for all samples. It is clear from Fig. 10 that AC conductivity increases with increasing frequency. As discussed in the dielectric constant characteristics, the grains boundaries are more effective than grains in the low frequency region. It means that the conduction mechanisms are controlled by the volume of grain boundaries in the low frequency region. The motion of charge carries is affected by the grain boundaries in the lower frequency region due to the low conductivity. Due to the higher resistance of grain boundaries, the grain boundaries act as barriers to the conduction of material. When the applied frequency equals the hopping frequency of electrons the dielectric behaviour of the material breaks. At higher frequencies dielectric constant of the material is becoming constant. As per eq. (6), if ε' is constant and variation in tangent (except $x = 0.2$) loss is small then conductivity depends directly on frequency. Therefore the AC conductivity increases with the increase in frequency. The maximum value of AC conductivity is observed for $x = 0.2$. The presence of loss peak at higher frequencies for x = 0.2 may be responsible for high values because of the increase in jumping frequency of electrons. The AC conductivity increases in the high frequency region attributed to the contribution of grains.

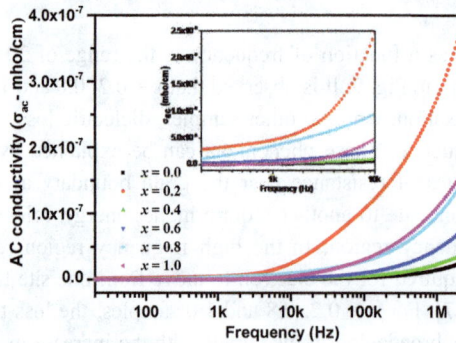

Fig. 10. Frequency dependent AC conductivity (σ_{ac}) of $SrCo_xAl_xFe_{12-2x}O_{19}$ ($0.0 \leq x \leq 1.0$) samples heated at $650°C$ for 3 h.

Dielectric Modulus (M') and (M'') were calculated using equations (7) and (8) respectively,

$$(M') = \frac{\varepsilon'(\omega)}{[\varepsilon'(\omega)]^2 + [\varepsilon''(\omega)]^2} \tag{7}$$

$$(M'') = \frac{\varepsilon''(\omega)}{\left[\varepsilon'(\omega)\right]^2 + \left[\varepsilon''(\omega)\right]^2} \qquad (8)$$

Fig. 11(a) shows the variation of real part of the modulus (M') with frequency for all the samples at room temperature. Modulus spectroscopy is an important and useful tool for determining, analyzing and interpreting the relaxation phenomena occurring in electrically and ionic conducting materials [39]. The real part of dielectric modulus is calculated using the formula mentioned in reference [40]. It is evident from Fig. 11 (a) that M' has smaller values in the lower frequency region and it increases with the frequency acquiring a maximum value at 2 MHz for all the samples.

The frequency dependence of the imaginary part of the complex modulus (M'') is shown in Fig. 11 (b). A broad peak is observed for all concentration indicating the existence of relaxation mechanism for the hexagonal ferrites [41]. The frequencies representing maximum peak are responsible for the mobility of charge carriers. At the frequencies above the maximum peaks, the charge carriers are confined to potential well and are immobile [42].

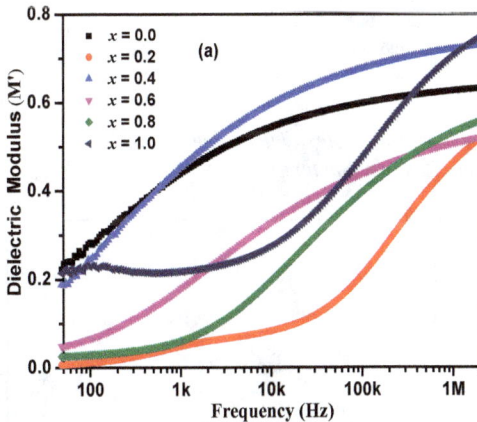

Fig. 11 (a). Real part of complex dielectric modulus (M') of SrCo$_x$Al$_x$Fe$_{12-2x}$O$_{19}$ (0.0 ≤ x ≤ 1.0) samples heated at 650° C for 3 h.

Cole-Cole plots (M'' vs M') were studied (Fig. 12) to examine the role of grain and grain boundaries in the prepared hexagonal ferrites. Cole-Cole plot shows only one semi-circle for all the compositions. The left side of the semicircle towards the lower frequency region contributes to the grain resistance [43] whereas the intermediate frequency under the curve represents grain boundary contribution [44]. The portion of the extreme right towards the higher frequency region represents resistance for the grains and grain boundaries [45]. It is clear from the figure that substitution makes a small difference in the grain resistance except for the sample $x = 1.0$. Substitution increases the grain boundary resistance.

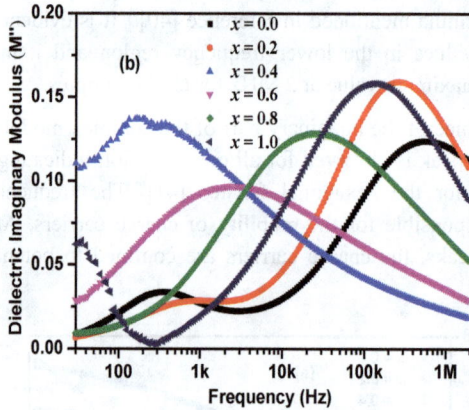

Fig. 11 (b). Imaginary part of complex dielectric modulus (M'') of $SrCo_xAl_xFe_{12-2x}O_{19}$ (0.0 ≤ x ≤ 1.0) hexagonal ferrites at room temperature.

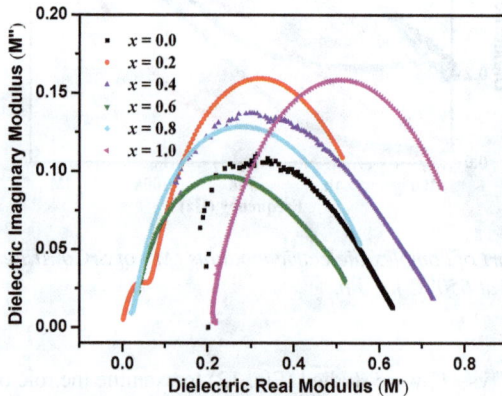

Fig. 12. Cole-Cole plots of $SrCo_xAl_xFe_{12-2x}O_{19}$ (0.0 ≤ x ≤ 1.0) samples heated at 650° C for 3 h.

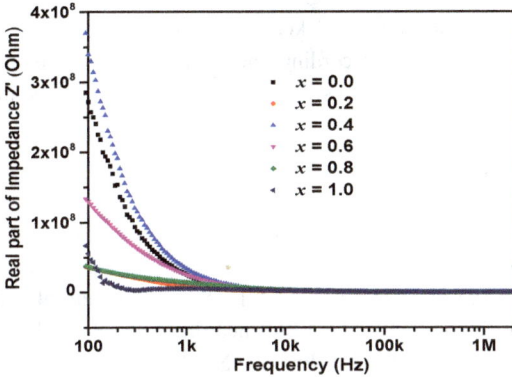

Fig. 13. Frequency dependant real part of impedance (Z') of $SrCo_xAl_xFe_{12-2x}O_{19}$ (0.0 ≤ x ≤ 1.0) samples heated at 650° C for 3 h.

The variation of the real part of dielectric impedance (Z') with frequency is shown in Fig. 13. It is observed that the real part of impedance is higher at the lower frequency and dispersion in real impedance is observed with the increase in the frequency and then a straight line is seen after the frequency of 2.5 KHz. At higher frequencies,all curves merge by decreasing the value of Z' to zero demonstrating non dependency of Z' on frequency. The reduction in the space charge polarization leads to the merging of curves for all the samples [45].

The faster recombination of space charge due to their smaller time to relax at higher frequency region results in the merging of curves. It is also observed that increase in Co-Al content, decreases the real value of dielectric impedance. The decrease in the value of Z' with the increase in frequency indicates the increase in conductivity at a higher frequency. This result can be compared with the results of AC conductivity.

Conclusions

A series of $SrCo_xAl_xFe_{12-2x}O_{19}$ (0.0 ≤ x ≤ 1.0) hexaferrites were synthesized using a simple heat treatment method. The effect of cobalt and aluminum substitution on structural, magnetic and dielectric properties of strontium hexaferrite is reported. XRD analysis shows the formation of mixed phases of M-type hexagonal ferrite and α-Fe_2O_3. The samples with x < 0.4 showed high coercivity compared to typical values of ~ 4 kOe for SrM. This high coercivity, with moderate remanence of > 23.96 emu/g makes these compounds suitable for permanent magnet applications. At higher concentrations ($x >$ 0.8) the coercivity drops to below 2.5 kOe which makes the compounds suitable for potential applications of magnetic recording.The average particle size ranges from 24 nm to 46 nm. The dielectric constant is observed to be decreasing with the increase in Co-Al substitution.

Acknowledgements

One of the authors Chetna C. Chauhan is thankful to Gujarat Council of Scientific Research (GUJCOST), Gandhinagar, for financial support in the form of the project (GUJCOST letter No. GUJCOST/MRP/14-15/1119). Authors are also thankful to CIF, IIT Guwahati, India, for providing the VSM measurements. R. B. Jotania thanks DST, UGC, India for DST-FIST, DRS-SAP (Phase I, No. UGC-F.530/10/DRS/2010 (SAP-I)) grants.

References

[1]   R. C. Pullar, Hexagonal ferrites: a review of the synthesis, properties and applications of hexaferrite ceramics, Progress in Materials Science, 57 (2012) 1191-1334.

[2]   S. R. Shinde, S. E. Loland, C. S. Ganpule, S. M. Bhagat, S. B. Bhagat, S. B. Ogle, R. Ramesh, T. Venkatesan, Improvement in spin-wave resonance characteristics of epitaxial barium-ferrite thin films by using an aluminium doped strontium-ferrite buffer layer, Applied Physics Letters, 74 (4) (1999) 594-596.

[3]   N. Chen, K. Yang, M. Y. Gu, Microwave absorption properties of La substituted M-type strontium ferrites, Journal of Alloys and Compounds, 490 (2010) 609-612.

[4]   J. Dho, E. K. Lee, J. Y. Park, N. H. Hur, Effects of the grain boundary on the coercivity of barium ferrite $BaFe_{12}O_{19}$, Journal of Magnetism and Magnetic Materials 285(1-2) (2005) 164-168.

[5]   B. K. Rai, S. R.Mishra, V. V. Nguyen, J. P. Liu, Synthesis and characterization of high coercivity rare-earth doped $Sr_{0.9}RE_{0.1}Fe_{10}Al_2O_{19}$ (RE: Y, La, Ce, Pr, Nd, Sm,and Gd), Journal of Alloys and Compounds,550 (2013) 198-203.

[6]   Y. Xu, G. L. Yang, A. P. Chu, H.R. Zhai, Theory of the single ion magnetocrystalline anisotropy of 3d ions, Physica Status Solidi B, 157 (1990) 685-693.

[7]   C. A. Van Den Brock, A. L. Stuijts, Ferroxdure, Philips Technical Review, 37(7) (1977) 157-175.

[8]   O. Kubo, T. Ido, H. Yok, Properties of Ba ferrite particles for perpendicular magnetic recording media, IEEE Transactions on Magnetics, 18(6) (1982) 1122-1124.

[9]   S. Hussain, N. Abbas Shah, A. Maqsood, A. Ali, M. Naseemand W. Ahmad Adil Syed, Characterization of Pb doped Sr ferrites at room temperature, Journal of Super Conductivity and Novel Magnetism, 24 (2011)1245-1248.

[10] W. Zhong, W. P. Ding, N. Zhang, Key Step in the synthesis of ultrafine $BaFe_{12}O_{19}$ by sol-gel synthesis, Journal of Magnetism and Magnetic Materials, 168(1-2) (1997) 196-202.

[11] A. Drmota, M. Drofenik, A. Znidarsic, Synthesis and characterization of nano-crystalline strontium hexaferrite using the coprecipitation and microemulsion methods with nitrate precursors, Ceramics International, 38 (2012) 973-979.

[12] W. Onreabroy, K. Papato, G. Rujijanagul, K. Pengpat, T. Tunkasiri, Study of strontium ferrites substituted by lanthanum on the structural and magnetic properties, Ceramics International, 38S (2012) S415-S419.

[13] R. Martinez Garcia, V. Bilovol, L. M. Socolovsky, Effect of the heat treatment conditions on the synthesis of Sr-hexaferrite, Physica B: Condensed Matter, 407 (16) (2011) 3109-3112.

[14] Z. Ullah, S. Atiq, S. Naseem, Influence of Pb doping on structural, electrical and magnetic properties of Sr-hexaferrites, Journal of Alloys and Compounds, 555 (2013) 263-267.

[15] S. Ounnunkada, P. Winotai, Properties of Cr-substituted M-type barium ferrites prepared by nitrate-citrate gel-autocombustion process, Journal of Magnetism and Magnetic Materials, 301 (2007) 292-300.

[16] L. Q. You, J. Zheng, The magnetic properties of strontium hexaferrite with La-Cu substitution prepared by SHS method, Journal of Magnetism and Magnetic Materials, 318 (2007) 74-78

[17] Vinod Dhage, M. I. Mane, A. P. Keche, C. T. Birajdar, K. M. Jadhav, The structural and magnetic behaviorofaluminum doped barium hexaferrite nanoparticles synthesized by solution combustion technique, Physica B: Condensed Matter, 406(4) (2011) 789-793.

[18] M. R. Eraky, A. A. Beslepkin, S. P. Kuntsevich, Magnetic properties and NMR studies of the SrAl-M hexagonal ferrite system, Materials Letters, 57(2003) 3427-3430.

[19] J. N. Dahal, L. Wang, S. R. Mishra, V.V. Nguyen, and J.P. Liu, Synthesis and magnetic properties of $SrFe_{12-x-y}Al_xCo_yO_{19}$ nanocomposites prepared via autocombustion technique, Journal of Alloys and Compounds 595 (2014) 213-220.

[20] Jasbir Singh, Charanjeet Singh, Dalveer Kaur, S. Bindra Narang, Rajat Joshi, Sanjay R. Mishra, Rajshree Jotania, Madhav Ghimire, Chetna C. Chauhan, Tunable microwave absorption in Co-Al substituted M-type Ba-Sr hexagonal ferrite, Materials and Design, 110 (2016) 749-761.

[21] H. Luo, B. K. Rai, S. R. Mishra, V. V. Nguyen, J.P. Liu, Physical and Magnetic properties of highly aluminum doped strontium ferrite nanoparticles prepared by auto-combustion route, Journal of Magnetism and Magnetic Materials, 324 (17) (2012) 2602-2608.

[22] P. Priyadharsini, A. Pradeep, P. Sambasiva Rao, G. Chandrasekaran, Structural, spectroscopic and magnetic studies of nanocrystalline Ni–Zn ferrites, Materials Chemistry and Physics, 116 (2009) 207-213.

[23] D. A. Skoog, Principles of Instrumental Analysis, Saunders Golden Sunburst Series, 1985.

[24] S. K. Chawla, R. K. Mudsainiyan, S. S. Meena, S. M.Yusuf, Sol-gel synthesis, structuraland magnetic properties of nanoscale M-type barium hexaferrite $BaCo_xZr_xFe_{(12-2x)}O_{19}$, Journal of Magnetism and Magnetic Materials, 350 (2014) 23-29.

[25] W. D. Kingery, H. K. Bowen, D. R. Uhlmann, Introduction to Ceramics, 2nd Edition, Wiley, Wiley, New York, 1976, pp. 25-87.

[26] A. Baniasadi, A. Hashemi, A. Nemati, M. A. Ghadikolaei and E. Paimozd, Microstructural evolution and mechanical properties of Ti–Zr beta titanium alloy after laser surface remelting, Journal of Alloys and Compounds, 583 (2014) 325-328.

[27] D. Ramimoghadam, S. Bagheri, S. B. A. Hamid, Stale mono-disperse nanomagnetic colloidal suspensions: an overview, Colloids and Surfaces B: Biointerfaces, 133 (2015) 388-411.

[28] Vivek Dixit, Chandani N. Nandadasa, Seong-Gon Kim, Sungho Kim, Jihoon Park, Yang-Ki Hong, Laalitha S. I. Liyanage, Amitava Moitra, Site occupancy and magnetic properties of Al-substituted M-type strontium hexaferrite, Journal of Applied Physics, 117 (2015) 243904-243917

[29] D.G. Agresti, T.D. Shelfer, Y. K. Hong, Y. J. Paig, A Mössbauer study of CoMo-substituted barium ferrite, IEEE Transactions Magnetics, 34 (4) (1989) 4069-4071.

[30] C. Singh, S. B. Narang, I. S. Huduara, Y. Bai, C. Singh, S. B. Narang, I. S. Huduara, Y. Bai, Dielectric properties of lanthanum substituted barium titanate microwave ceramics, Journal of Alloys and Compounds, 464 (1) (2008) 429-433.

[31] I. Ali, M. Islam, M. N. Ashiq, M. A. Iqbal, M. Awan, S. Naseem, Role of Tb–Mn substitution on the magnetic properties of Y-type hexaferrite, Journal of Alloys and Compounds, 599 (2014) 131-138.

[32] Y. Li, R. Liu, Z. Zhang, C. Xiong, Synthesis and characterization of nanocrystalline $BaFe_{9.6}Co_{0.8}Ti_{0.8}M_{0.8}O_{19}$ particles, Materials Chemistry and Physics, 64 (2000) 256-259.

[33] A Kumar, Annveer, M. Arora, M. S. Yadav, R. P. Panta, Induced Size Effect on Ni Doped Nickel Zinc Ferrite Nanoparticles, Physics Procedia, 9 (2010) 20-23.

[34] L. Shrideshmukh, K. K. Kumar, S. B. Laxman, A. R. Krishna, G. Sathaiah, Dielectric Properties and Electrical Conduction in Yttrium Iron Garnet (YIG), Bulletin of Materials Science, 21(3) (1998) 219-226.

[35] C. G. Koops, On the dispersion of resistivity and dielectric constant of some semiconductors at audiofrequencies, Physical Review, 83 (1951) 121-124.

[36] P. Kuruva, P. Reddy Matli, B. Mohammad, S. Reddigari, S. Katlakunta, Effect of Ni-Zr co-doping on dielectric and magnetic properties of $SrFe_{12}O_{19}$ via sol-gel route, Journal of Magnetism and Magnetic Materials, 382 (2015) 172-178.

[37] W. D. Kingery, H. K. Bowen, D. R. Uhlmann, Introduction to Ceramics, 2nded., John Wiley & Sons, New York, 1976.

[38] V.R. K. Murthy, J. Sobhanadri, Dielectric properties of some nickel-zinc ferrites at radio frequency, Physica Status Solidi A, 36(2) (1976) K133-K135.

[39] R. Pandit, K.K. Sharma, P. Kaur, R.K. Kotnala, J. Shah, R. Kumar, Effect of Al^{3+} substitution on structural, cation distribution, electrical and magnetic properties of $CoFe_2O_4$, Journal of Physics and Chemistry of Solids, 75 (2014) 558-569.

[40] B.V. R. Chowdari, R.G. Krishnan, AC conductivity analysis of glassy silver iodomolybdate system, Solid State Ionics 23 (1987) 225–233.

[41] M. Irfan, A. Elahi, A Shakoor, Hysteresis and Electric modulus analysis of Y^{3+} doped MnNi –Y –type hexagonal ferrite, Ceramics Silikaty, 60(1) (2016) 34-40.

[42] S. M. Patange, S. E. Shirsath, K. S. Lohar, S. S. Jadhav, N. Kulkarni, K. M. Jadhav, Electrical and switching properties of $NiAl_xFe_{2-x}O4$ ferrites synthesized by a chemical method, Physica B: Condensed Matter, 406(3) (2011) 663-668.

[43] Yang. Bai, Ji. Zhou, Zhilun. Gui, Longtu. Li, Phase transformation, structure and magnetic properties of (Nd, Pr)–Fe–V–B alloys and their nitrides prepared by mechanical alloying, Journal of Magnetism and Magnetic Materials, 278 (2004) 208-213.

[44] M. G. Chourashiya, J. Y. Patil, S. H. Pawar, L. D. Jadhav, Studies on structural, morphological and electrical properties of $Ce_{1-x}Gd_xO_{2-(x/2)}$, Materials Chemistry and Physics, 109 (2008) 39-44.

[45] O. Raymond, R. Font, J. Portelles, N. Suarez-Almodovar, J. M. Siqueiros, Frequency temperature response of ferroelectromagnetic $Pb(Fe_{1/2}Nb_{1/2})O_3$ ceramics obtained by different precursors, Journal of Applied Physics, 99 (2006) 124101(1)-124101(9).

Chapter 9

Soft Ferrite: A Brief Review on Structural, Magnetic Behavior of Nanosize Spinel Ferrites

N.N. Sarkar[1,a], K.G. Rewatkar[1,b], V.M. Nanoti[2,c], N.T. Tayade[3,d]

[1]Department of Physics, Dr. Ambedkar College, Deeksha Bhoomi, Nagpur 440 010, India

[2]Department of Physics, Priyadarshini Institute of Engineering and Technology, Nagpur 440 019, India

[3]Department of Physics, Government Institute of Science, R. T. Road, Civil Lines, Nagpur 440008, India

[a]ns.rathinonly4u@gmail.com, [b]kgrewatkar@gmail.com, [c]viveknanoti@gmail.com, [d]nishanttayade@rediffmail.com

Abstract

In the present work, we have focused on crystal structure, magnetic properties, M-H curve, superparamagnetic behavior and magneto-crystalline anisotropy of soft ferrimagnetic oxides having spinel structure. Due to their unique magnetic properties such as high Curie temperature and high saturation magnetization, spinel ferrites offer opportunities for applications in different fields such as hyperthermia, targeted drug delivery, magnetic resonance, microstrip antenna. Various applications of these ferrites have been discussed and explored due to possibility nano-size material synthesis.

Keywords

Soft Ferrite, Spinel Structure, XRD, Magnetocrystalline Anisotropy, Biomedical Application

Contents

1. Introduction

Magnetism is very common to everything one sees and the term 'magnetism' comes from a rock called 'lodestone'. The name 'magnet' was first used by the Greeks for this lodestone, because of its property of attracting other pieces of the same fabric and iron as well. It was shown later that this naturally occurring lodestone is the magnetic iron oxide or the naturally occurring mineral called 'magnetite' having chemical composition Fe_3O_4 [1]. It is important to understand that the large variety of magnetic materials and their attributes are mainly related to three ferromagnetic elements like iron, cobalt, and nickel. Ferrites are a group of ferrimagnetic materials [2]. Ferrite is a ceramic-based ferrimagnetic material, including iron oxides and complex iron oxides containing other metals such as a rare earth or an alkali earth. Ferrites show both the lower and higher saturation magnetization depend upon soft and hard magnetic materials, they exhibit several advantages, including applicability at higher frequencies, lower electrical conductivity, higher heat as well as greater corrosion resistance, low eddy current and dielectric losses. No material with such wide ranging properties exists and therefore ferrites are unique magnetic materials which find applications in almost all area of life. Basically, ferrites are of four types, namely spinel, garnet, hexaferrite and orthoferrite. All these ferrites have different crystal structure, chemical composition and all are equally important for technological applications [3]. Spinel ferrites are represented by the formula unit $MeFe_2O_4$ (where Me represents divalent cations). Most of the spinel ferrites possess a cubic crystal structure with oxygen anions in FCC positions and cations on the tetrahedral (A) and octahedral (B) coordinated interstitial lattice sites, form the A and B sub-lattices [4]. For example 6 times of Fe_2O_3 prototype is being used in forming the hexaferrites which has a hexagonal crystal structure with general formula $MeFe_{12}O_{19}$ (where Me= Ba, Sr, Ca etc.) Orthoferrites crystallize in orthorhombic distorted perovskite structure having space group Pbnm. Garnets have a complex cubic structure having general formula $R_3Fe_5O_{12}$ (where R= Yttrium or rare earth ions like Dy, Gd, La etc.) Since the proposed review is based on spinel ferrites which have a cubic symmetry; the crystal chemistry of spinel ferrites is covered in the next section [5, 6].

2. Crystal structure and cation distributions

2.1 Structure of spinel ferrites

Compounds with the cubic spinel structure have the general formula $MeFe_2O_4$. The anions O^{-2} form a cubic close packed lattice. The 'Me' ions occupied at tetrahedral interstices and the 'Fe' ions at octahedral site [7]. Fig.1(a) shows the unit cell of spinel ferrites consists of 32 oxygen anions, 16 trivalent iron ions, and 8 divalent metal ions. The placement of metal cations (16 trivalent + 8 divalent = 24) in a unit cell is shown in Fig.1 (b). The most significant characteristic of the unit cell is that its array of oxygen ions reveals two types of interstices (referred as tetrahedral or 'A' sites and octahedral or 'B' sites), which are occupied by the metal ions as shown in Fig. 1(c) [8]. The unit cell possesses 32 oxygen ions, along with 64 tetrahedral (A) sites and 32 octahedral (B) sites and these sites are filled with metal ions, of either 2^+ or 3^+ valence ions; the positive charge would be much greater than the negative charge and hence the structure is not electrically neutral. If, as in the mineral, spinel, the tetrahedral sites are occupied by divalent ions and the octahedral sites are occupied by the trivalent ions, the total positive charge would be $8 \times (2^+) = 16^+$; the $16 \times (3^+) = 48^+$ or a total of 64^+ which is needed to balance the $32 \times (2^-) = 64^-$ for the oxygen ions. There would then be eight formula units of $MeO \cdot Fe_2O_3$ or $MeFe_2O_4$ in a unit cell to neutralize the ionic charge balance.

2.2 Interstitial sites in spinel structure

In a spinel structure, large oxygen ions form a FCC lattice, the cubic close packed structure have two kinds of interstitial sites namely, tetrahedral (A) and octahedral (B) sites which are surrounded by four and six oxygen ions respectively. In a tetrahedral site, interstitial is situated in the middle of a tetrahedron formed by four lattice atoms. Out of 4 lattice atoms three atoms are touching each and exist in the same plane, the fourth atom occurs in the symmetrical position on top as shown in Fig. 2(a). Therefore, the tetrahedral site has a defined geometry and offers space for an interstitial atom. On the other hand, octahedral position for an atom is a space in the interstices between 6 regular atoms that form an octahedron, four regular atoms are positioned in a plane and other two are in a symmetrical position just above or below the center atom shown in Fig. 2(b).

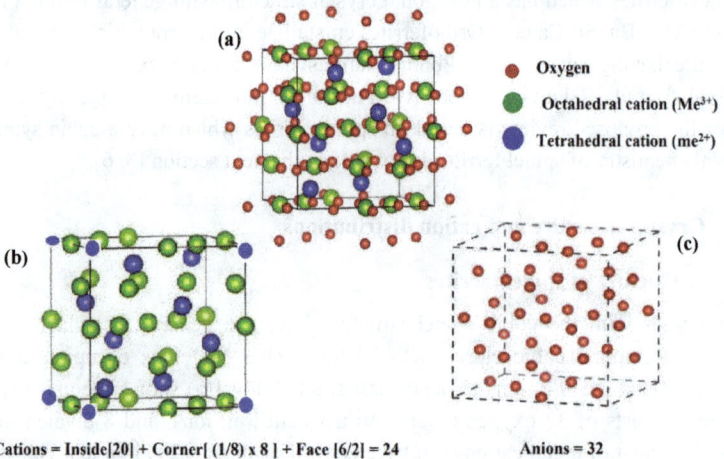

Number of Cations = Inside[20] + Corner[(1/8) x 8] + Face [6/2] = 24   Anions = 32

Fig. 1(a–c). Unit cell of spinel structure (a) and distribution of cations (b) and anions (c).

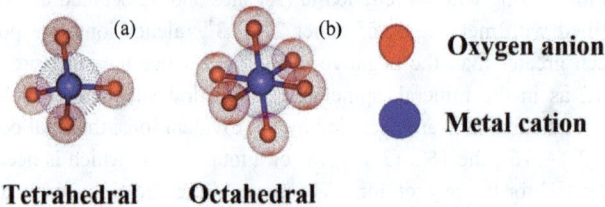

Fig. 2(a, b). Tetrahedral (A) and Octahedral (B) sites.

2.3 Types of spinel

2.3.1 Normal spinel

On the basis of compositional variation of magnetic ferrites $Me^{2+}Fe_2^{3+}O_4$ spinel structure can either have a 'normal' structure, in which the divalent ions (Me^{2+}) are sat on tetrahedral sites (Me^{2+}), where as Fe^{3+} ions occupy on octahedral sites. The general formula for normal spinel is (Me^{2+}) [Fe^{3+}]$_2O_4$, square brackets indicate B sites occupancy. The distribution of normal spinel is shown in Fig. 3(a) [9].

2.3.2 Inverse spinel

In an inverse spinel divalent cations are on B sites, where as trivalent cations are equally distributed among the A and B sites. For example Fe_3O_4, in which half of its Fe^{3+} ions (or Me^{3+}) are located on A sites while other half of Fe^{2+} ions occupies the B sites. The formula for an inverse spinel is (Fe^{3+}) [$Me^{2+}Fe^{3+}$]O_4 as shown in Fig. 3(b) [10].

2.3.3 Mixed spinel

In some cases, the divalent cation has no specific site preference among A or B site, so that it circulates in both sites leading to a formation of mixed spinel (Fig. 3(c)). The general formula for cations distribution among A and B sites can be represented as ($Me_{1-\delta}^{2+}$ Fe_δ^{3+})^A[Me_δ^{2+} $Fe_{2-\delta}^{3+}$]$^BO_4$, where 'δ' is the degree of inversion. The degree of inversion depends mainly on the method of preparation and synthesis conditions, time etc. [11, 12]. The migration of Fe^{3+} cations among tetrahedral and octahedral sites of the spinel mentioned above has a peculiar physico-chemical property. For example, inverse spinel tends to be ferrimagnetic, whereas normal spinel have paramagnetic nature [13].

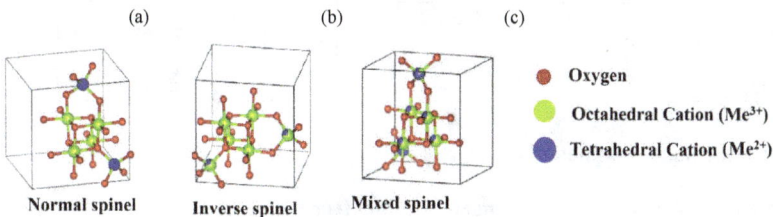

Fig. 3(a–c). Cations distribution among normal, inverse and mixed spinels.

2.4 Cation distributions of spinel ferrites

The importance of the distribution of the cations over A and B sites for the magnetic properties can be demonstrated by the differences between $NiFe_2O_4$, $ZnFe_2O_4$, $MgFe_2O_4$ etc. all compound contain trivalent Fe and a non-magnetic divalent ion, Zn^{2+}, Mg^{2+} Ni^{2+}

etc. as shown in Table 1. The saturation magnetization in spinels depends upon the cation distributions, particle size, annealing temperature and time etc. The magnetic behavior and net magnetic moment of few spinel ferrites is indicated in Table 1. The magnetic order in cubic spinels is mainly due to super exchange interactions among the metal ions and sublattices. Thus, it is possible to vary magnetic properties of the samples spinels by varying the cations. Table 1 suggests that Cobalt and Chromium ions predominantly occupy the octahedral sites, which is coherent with their predilection for large octahedral site energy. Nevertheless, in the case of Co-Cr ferrites saturation magnetization (M_s) decreases, because all the iron ions prefer the tetrahedral (A) site. The decrease in Ms in Co-Cr ferrites are due to the B-B interaction and are more dominating than A-B interaction.

Table 1: Cation distributions in various spinel ferrites.

Spinel ferrite	Structure	Tetrahedral (A) site	Octahedral (B) site
$(Fe)^A [NiFe]^B O_4$	Inverse	Fe^{3+}	Ni^{2+},Fe^{3+}
$(Zn)^A [Fe_2]^B O_4$	Normal	Zn^{2+}	Fe^{3+}
$(Fe_{0.2}Mg_{0.8})^A [Mg_{0.2}Fe_{1.8}]^B O_4$	Partially Inverse	Fe^{3+}	Mg^{2+},Fe^{3+}
$(Fe)^A [CoCr]^B O_4$	Inverse	Fe^{3+}	Cr^{3+},Co^{2+}
$(Fe_{0.8}Cr_{0.2})^A [NiCr_{0.8}Fe_{0.2}]^B O_4$	Mostly inverse	Fe^{3+}, Cr^{3+}	Ni^{2+},Fe^{3+},Cr^{3+}

3. Structural and magnetic behavior

3.1 X-ray diffrection technique used in spinel ferrites

The XRD technique is the most common tool used to identify the crystal structure of a sample as depicted in JCPDS file, mostly researchers used to match the specific (hkl) planes such as (111) (220) (311) (222) (440) (422) (511) (400) in spinel ferrites with their XRD data. These (hkl) planes are major reflections observed for the sample at an appropriate 2θ position [14]. On the other hand, if divalent, trivalent and tetravalent cations are substituted in the stoichiometric ratio of $MeFe_2O_4$ of each formula unit, then the normal spinel structure is owing to change as an inverse spinel as well as mixed spinel ferrite [15]. In general spinel ferrite belongs to the Space group Fd-3m, with a lattice constant a = 8.000 Å (considering error up to ± 0.5 Å). In the present research work we have synthesized $Mg_{(1+x)}Zr_x(FeCr)_{1-x}O_4$ where x = 0.2, 0.4 and characterized through XRD and VSM (Vibrating Sample magnetometer) to understand its structural and magnetic behavior. The sample was prepared by the sol-gel auto combustion method using magnesium nitrate, ferric nitrate, chromium nitrate, zirconium oxynitrate and urea in a proper molar ratio. Further the XRD data can be refined through Rietveld analysis as shown in Fig. 4. From the Rietveld refinement, Bragg R factor and the values of χ^2 have been calculated and listed in Table 2. Further, the XRD data converted into CIF file and was analyzed using VESTA software, and the output image as shown in Fig. 5. If the goodness of fitting (χ^2) is in between 1−8 then it is considered to be reliable. The particle size has been calculated using Debye-Scherrer formula given below

$$D_{XRD} = \frac{(0.9\,\lambda)}{\beta \cos\theta} \tag{1}$$

Where 'λ' is the wavelength of Cu-k_α radiation, 'β' is the full width half maximum of diffraction peak, and 'θ' is the Bragg angle. The lattice parameter (a) was determined using the relation:

$$a = d_{hkl}(h^2 + k^2 + l^2)^{\frac{1}{2}} \tag{2}$$

Where 'h', 'k' and 'l' are miller indices of plane. The lattice parameter (a) and crystallite size (D_{XRD}) are summarized in Table 2.

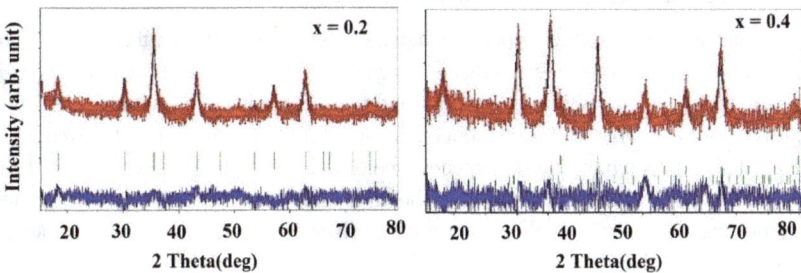

Fig. 4. Rietveld refinement XRD patterns of $Mg_{(1+x)}Zr_x(FeCr)_{1-x}O_4$ (x = 0.2, 0.4).

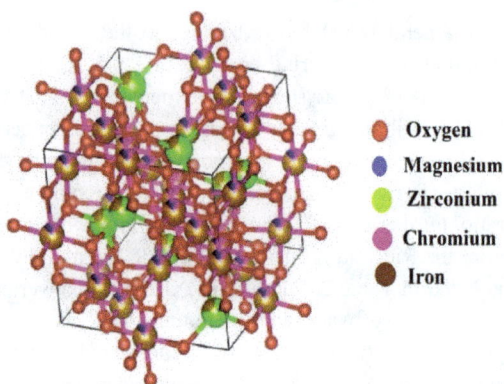

Fig. 5. Distribution of cations in octahedral and tetrahedral site.

Table 2: Lattice parameter and average crystalline size of $Mg_{(1+x)}Zr_x(Fe\ Cr)_{1-x}O_4$.

Sr. No	Compositions	Lattice parameter ($a = b = c$) (Å)	Average crystallite size (D_{XRD}) (nm)	Bragg R- Factor	χ^2
1	$Mg_{1.2}\ Zr_{0.2}\ (Fe\ Cr)_{0.8}\ O_4$	8.3341	15	23.5943	5.06
2	$Mg_{1.4}\ Zr_{0.4}\ (Fe\ Cr)_{0.6}\ O_4$	8.4122	20	49.8211	6.10

3.2   Interactions among magnetic moments on lattice sites

In spinel ferrites there are three types of magnetic interactions between the metallic ions mediated by oxygen anion (super–exchange mechanism), known as A–A interaction, B-B interaction and A–B interaction. It has been observed experimentally that the interaction energies are negative [16], therefore inducing an anti-parallel orientation. The strength of super exchange interaction energy depends upon angle between Me^{2+}–O–Fe^{3+} and the distance from these ions to the oxygen as shown in Fig. 6. The distance between cation and anion site (Me^{2+}– O / Fe^{3+} – O) and the distance between Me^{2+}– Fe^{3+} can be calculated by using the formula given in Table 3.

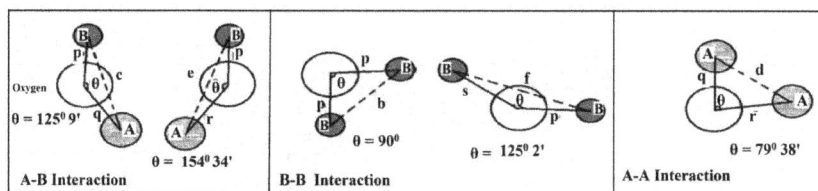

Fig. 6. Magnetic interactions among A–A, A–B and B–B sites.

Based on the values of the distance and the angle θ, among the three interactions the A-B interaction is of the highest magnitude. There are two configurations for A–B interaction has small distances (p, q and p, r) and the values of the angle θ are higher than all other. In case of B–B interaction there are configurations but only the first one will be effective because the distance 's' is too large for second configuration The A–A interaction is the weakest, as the distance r is large and the angle θ is small

Table 3: Formula for Metal–Oxygen and Metal–Metal interactions.

Metal–Oxygen bonds $Me^{2+}- O / Fe^{3+} - O$	Metal–Metal bonds $Me^{2+} – Fe^{3+} / Me^{2+} – Me^{2+} / Fe^{3+} – Fe^{3+}$
$P = a\left(\dfrac{1}{2} - u\right)$	$b = \left(\dfrac{a}{4}\right) 2^{\frac{1}{2}}$
$q = a\left(u - \dfrac{1}{8}\right) 3^{\frac{1}{2}}$	$c = \left(\dfrac{a}{8}\right) 11^{\frac{1}{2}}$
$r = a\left(u - \dfrac{1}{8}\right) 11^{\frac{1}{2}}$	$d = \left(\dfrac{a}{4}\right) 3^{\frac{1}{2}}$
$s = \dfrac{a}{3\left(u - \dfrac{1}{2}\right) 3^{\frac{1}{2}}}$	$e = \left(\dfrac{3a}{8}\right) 3^{\frac{1}{2}}$
Where a = Lattice parameter and u = Anions position parameter	$f = \left(\dfrac{a}{4}\right) 6^{\frac{1}{2}}$

3.3 Magnetic moments of spinels

Crystal Field Stabilization Energy (CFSE) in the theory of crystal field, gives an idea about the distribution of electrons in the d orbital's which may lead to net stabilization (decrease in energy) of some complexes depending on the specific ligands field geometry and metal d– electron configurations. When the transition metal ions are surrounded by ligands, then the d– orbital of Me^{2+} and Me^{3+}ions splits in to lower and higher energy level as shown in Fig. 7.

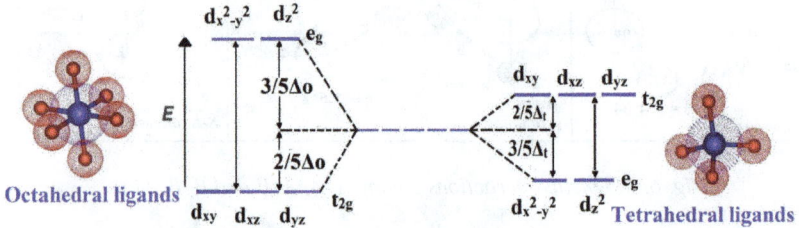

Fig. 7. Splitting of energy levels in spinel.

In the present case spinel ferrite having all the 'd' orbital of metal ions either completely filled or a maximum of five unpaired electrons. These unpaired electrons having its individual magnetic moments and hence the magnetization of the materials depends on the magnitude of the magnetic moments existing in the system. There are two possibilities of d orbital splitting: (1) due to octahedral ligands field, and (2) due to tetrahedral ligands field. The energy gap between lower and higher energy is denoted by Δ_o and Δ_t (Δ_o due to the field of octahedral ligands and Δ_t due to tetrahedral ligands) as shown in Fig. 8. There are five d orbits, referred to as d_{xy}, d_{xz}, d_{yz}, d_z^2 and $d_{x^2-y^2}$. For an octahedral complex, an electron in the more stable t_{2g} subset can be treated as contributing $-2/5\Delta_o$, whereas an electron in the higher energy e_g subset contributes to a destabilization of $+3/5\Delta_o$. In case of tetrahedral ligands Δ_t is less than Δ_o because of its weak field therefore in a tetrahedral complex, an electron is more stable t_{2g} subset contributing $+2/5\Delta_t$ whereas an electron in the lower energy e_g subset contributes to a destabilization of $-3/5\Delta_t$ as shown in Fig. 8.

There are two possibilities for metal ions having d^5 electronic configuration. Depending upon the nature of the ligands and the metal they could be high–spin or low–spin complexes shown in Fig. 9. If the field is weak then the electrons occupy both the energy states that produce higher spins, and if the field is strong then all the electrons occupy the lower energy state since it makes lower spins. The spin of electrons is directly proportional to magnetization of the materials, all the above phenomena is describe by crystal field theory therefore CFSE is one of the most important parameter to understand the structural and magnetic properties of spinel ferrite.

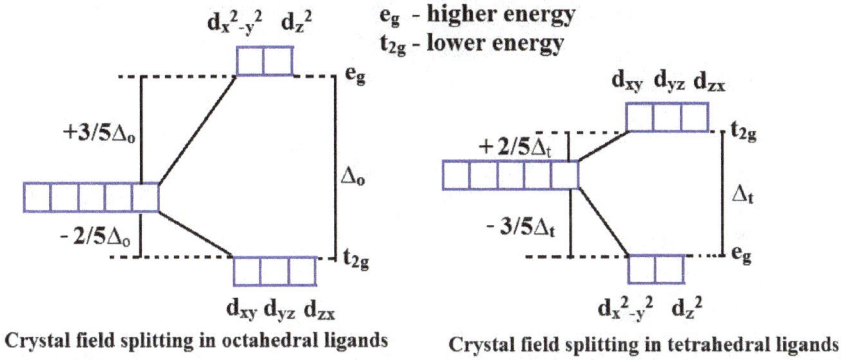

Crystal field splitting in octahedral ligands   **Crystal field splitting in tetrahedral ligands**

Fig. 8. Splitting mechanism.

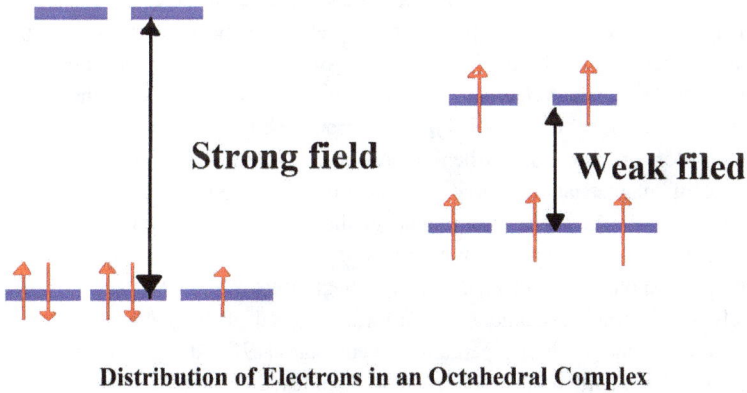

Distribution of Electrons in an Octahedral Complex

Fig. 9. Distribution of electrons in octahedral complex system.

For the d^5 system of high spin,

$$CFSE = (3 \times 0.4) - (2 \times 0.6) = 0 \; \Delta_o \tag{3}$$

For low spin,

$$CFSE = 1 \times 0.4 = 0.4 \; \Delta_o \tag{4}$$

3.4 Superparamagnetism

Magnetic material exhibits magnetic ordering, that can be observed by using M–H loop which reveals the magnetic nature of that material. Fig.10 describes the behavior of ferrimagnetic, paramagnetic and the super-paramagnetic substances. Super-paramagnetism (SPM) is a state of magnetism revealed by small ferrimagnetic nanoparticles of typical diameter ≤ 100 nm, depending upon the types of material. The total magnetic moment of the nanoparticles can be viewed as one giant magnetic moment, composed of all the individual magnetic moments of the atoms, and the nanoparticles are characterized by a modest value of coercivity, so they cause low hysteresis loss at high frequency [17]. Superparamagnetism (SPM) prevails in the range of a single–domain–sized grain when thermal energy is sufficient to surmount barriers to a reversal of magnetization. These barriers arise from magneto-crystalline and shape anisotropy, all of which are proportional to the grain volume [18]. When the energy barriers are larger than the thermal energy, the magnetization is "blocked" and the probability of spontaneous reversal becomes negligible. When the energy of barriers is relatively low, thermal excitations can result in reversal of the magnetization over very short timescales and are in a superparamagnetic state [19]. At a given temperature, the volume at which a particle moves from being unblocked to the blocked state is known as the blocking volume (V_b). For a given volume, unblocked particles become blocked as the temperature is lowered below the critical blocking temperature (T_b). Very often, nanoparticles show a certain preference for the orientation of the magnetic moment along which their magnetization aligns, in which case these nanoparticles are said to have magnetic anisotropy, and if it is mainly one preferred direction, then it is called uniaxial anisotropy. For uniaxial particles randomly oriented in the direction of their magnetization, which is caused by thermal energy,–the median relaxation time is calculated using the relation:

$$t = t_o \exp\left(\frac{\Delta E}{k_B T}\right) \tag{5}$$

Where t_o is the length of time characteristic of the probed material. Often it's of a magnitude of around 10^{-9} to 10^{-12} s, ΔE is the energy barrier in the magnetization, k_B is Boltzmann constant and T is the temperature. The observation of nanoparticles in a superparamagnetic state however does not just depend on the temperature T, and the energy barrier ΔE. Each experimental technique comes with its own measurement time (t_m). Depending on the measurement time, the following two scenarios can be observed. The 'Superparamagnetic property' is one of that exclusive attributes of magnetic nanoparticle, which is directly hooked on their magnetic anisotropy. In small size NPs, (Nano particles) magnetization can randomly flip direction under the influence of temperature. The typical time between two pitches is called Neel relaxation time. In the absence of an external magnetic field, when the time used to measure the magnetization of the NPs is much longer than the Neel relaxation time, their magnetization appears to be at zero, and the particles appear to be in the super-paramagnetic state. Possible reasons for imperfect superposition could be anisotropy effects and change of spontaneous magnetization with respect to temperature variation. In order to calculate the temperature dependence of the magnetic response of NPs, it is necessary to consider the effect of thermal fluctuations that allow a transition between magnetic states.

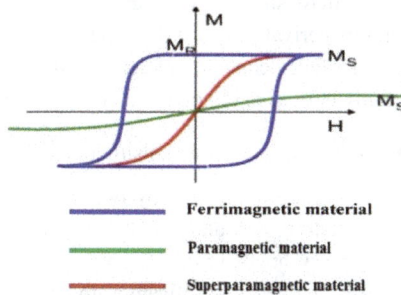

Fig. 10. M–H curves for Ferrimagnetic, Paramagnetic and Superparamagnetic materials [23].

It was observed that when a sample is cooled from a temperature where all particles show super-paramagnetic behavior to the lowest reachable temperature, then a small constant field is applied and the sample is heated to a temperature high enough, then it was found that initially growth magnetization and subsequent decrease in magnetization of the material [20-22].

3.5 Superparamagnetism in terms of biocompatibility

Due to superparamagnetic property, low toxicity and high biocompatibility, metallic element solid solution nanoparticles are the measure of sensible material for several medical applications like addressing targets, cellular medical aid, tissue repair, nano-biosensors, drug delivery, magnetic resonance imaging, and magnetic fluid hyperthermia [24, 25]. These applications need high magnetization values of NPs and size of around 100 nm with uniform physical and chemical properties. In addition, it is essential to see the potential toxicity of magnetic NPs for their undefeated application in nano drugs [26]. The surface coating of the nanoparticle is incredibly necessary for the determination of the fate of magnetic nanoparticle for in-vitro and in-vivo applications, while the magnetic nanoparticles (MNP), without coating show agglomeration and possess hydrophobic behavior; they have a large surface area to volume ratio. A correct surface coating permits MNP to spread uniformly in magnetic fluids and improves stability. Biocompatible coating like silicon, gold, polyethylene glycol, etc. [27, 28] is most explored in nano drugs. Amongst various materials, oxides are an excellent surface modifier due to its glorious biocompatibility, stability, nontoxicity, etc. conjugation with numerous useful teams. Three general ways for coating of nanoparticles with oxide shell includes Stober technique, micro silicon oxide and precipitated acid in an exceeding solution. Silicon oxide and its derivatives coated on the surface of magnetic nanoparticle might modify their particle surface behavior and apply with chemicals inert layer for the nanoparticle that is especially helpful in biological systems [29, 30].

3.6 Magnetic nanoparticles and magnetic anisotropy

Magnetic nanoparticles (MNPs) are dominated by two main features, finite-size effects (single-domain, multi-domain structures, and quantum confinement) and surface effects, which results from the symmetry breaking of the crystal structure at the surface of the particle [31]. The surface effect becomes significant as the particle size decreases because of the ratio of the number of surface atoms to the core atoms increases. It is easily proven that several magnetic properties such as magnetic anisotropy, magnetic moment per atom, Curie temperature, and the coercivity field of MNPs are dissimilar than those of a bulk material [32]. In most medical application, the preferred size of the nanoparticle is typically around 10–50 nm. In this range usually, a nanoparticle becomes a single magnetic domain (for minimization of its magnetic energy) and shows superparamagnetic behavior when the temperature is above a certain temperature called the freezing temperature. In the super-paramagnetic state, a nanoparticle possesses a large magnetic moment and behaves like a giant paramagnetic atom with a loyal reaction to applied magnetic fields with negligible remanence and coercivity.

Most of the magnetic materials contain some type of anisotropy that affects their magnetic behavior [33-34]. The most common types of anisotropy are (a) magneto-crystalline anisotropy (or magnetic anisotropy or crystalline anisotropy); (b) surface anisotropy; (c) shape anisotropy; (d) exchange anisotropy; and (e) induced anisotropy (for example, by stress). All these anisotropies have an influence on the magnetic properties to a certain extent. In nanoparticle, shape anisotropy and magneto-crystalline anisotropy are the most significant parameters. Magneto-crystalline anisotropy arises from the spin field interaction and energetically favors alignment of the magnetic moments along a specific crystallographic direction called the easy axis of the fabric. The magneto-crystalline anisotropy depends on the type of material, temperature, impurities and is independent of the sample size. But the magnetic anisotropy constant in Aluminum substituted Co ferrite (spinel) increases with crystallite size due to the increase of the volume contribution to the anisotropy reported by Kumar and Kar *et al.* [8].

Shape anisotropy causes the magnetization depends on the condition of the sample. The magnetization of a long, thin needle-shaped sample is easier along its long axis compared to that along any of its short axis. For the nanoparticle, shape anisotropy is the predominant form of anisotropy stress anisotropy implies that magnetization might change with emphasis. It was demonstrated that magnetic anisotropy changes when the surfaces are modified or adsorb different molecules [35]. This stands for that surface structure significantly influence the magnetic anisotropy. Hence, due to their large ratio of open bulk atoms, the surface anisotropy of the nanoparticle could be more significant than both the crystalline and shape anisotropy. The surface coating of the nanoparticle can have an influence on their magnetic anisotropies and hence on their magnetic properties [36].

As per the Kumar *et al.* in cobalt ferrite when Aluminum substitution increased to replace trivalent Fe^{+3} then the anisotropy decreased from 2.27×10^6 erg/cm^3 for $CoFe_2O_4$ to 1.40×10^6 erg/cm^3 for $CoFe_{1.6}Al_{0.4}O_4$, due to the decrease of spin–orbit coupling [8]. In the more complex case of $CuGa_xAl_xFe_{2-2x}O_4$, the effective magnetic anisotropy constant K_{eff} decreases on the substitution (x) of nonmagnetic ions in it [37].

3.7 Magnetic behavior of $Mg_{(1+x)}Zr_x(Fe\ Cr)_{1-x}O_4$

To understand the magnetic behavior of spinel ferrite for the present investigation, we have synthesized $Mg_{(1+x)}Zr_x(FeCr)_{1-x}O_4$ ($x = 0.2, 0.4$) using the sol –gel auto combustion method as discussed in the previous section. Fig. 11 represents field dependent magnetization (B–H) curves of Zr and Cr substituted Magnesium ferrites measured under an applied field of 15 KG.

It is noted from Fig. 11 that magnetization decreases with the increase in concentration of Zr^{4+} ions. The increasing coercivity is an indication of the change from superparamagnetic to ferrimagnetic nature, which is verified experimentally (Table 4). It is known that the total magnetization in ferrites depends on the anti–parallel magnetic interaction between the magnetic moments of cations, situated in tetrahedral (A-site) and octahedral (B-site) respectively, and the overall magnetization arises from the difference between the magnetization of the A and B sites, respectively. The Zr^{2+} occupied tetrahedral site therefore the magnetic moment of tetrahedral site is expected to decrease and as the radius of A-site cation increases, which upsize the average distance between A and B site cations. Therefore it augments the anti parallel interactions among A and B sites and hence the magnetic moments gradually declines. On the basis of above explanation, we may conclude that the magnetization step down with rising the value from $x = 0.2$ to $x = 0.4$ in $Mg_{(1+x)}Zr_x(FeCr)_{1-x}O_4$.

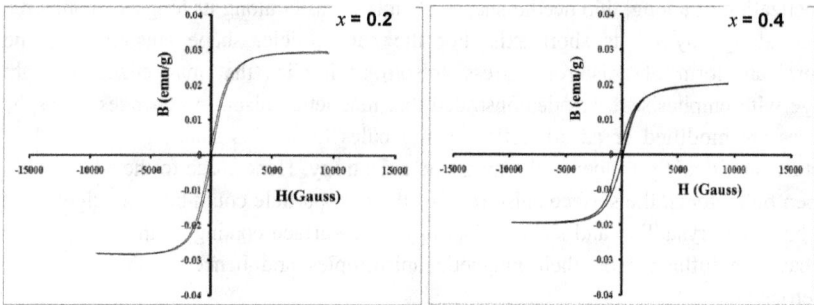

Fig. 11. B-H curves of $Mg_{(1+x)}Zr_x(FeCr)_{1-x}O_4$ ($x = 0.2, 0.4$).

Table 4. Magnetic parameters of $Mg_{(1+x)}Zr_x$ (Fe Cr)$_{1-x}O_4$ ($x = 0.2, 0.4$).

Samples	Saturation Magnetization (M_s) (emu/g)	Retentivity (M_r) (emu/g)	Coercivity (H_c) (Gauss)	Bohr magnetron (μ_B)/molecule
$Mg_{1.2}Zr_{0.2}(FeCr)_{0.8}O_4$	29.7×10^{-3}	1.89×10^{-3}	20	1.0477×10^{-3}
$Mg_{1.4}Zr_{0.4}(FeCr)_{0.6}O_4$	20×10^{-3}	1.5×10^{-3}	190	0.4843×10^{-3}

4. Applications of soft ferrites

4.1 Uses of spinel ferrites in physical devices

$CoFe_2O_4$, $NiFe_2O_4$, $(Mg, Mn)Fe_2O_4$ ferrites are important components for the digital products such as cellular phones, video cameras, note book computers and floppy drives, thermistors [38, 39] etc.

$Ni–Zn–CuFe_2O_4$ ferrites are explored in production of multilayer type chips, mainly because these oxides can be sintered at relatively low temperature. Due to lower densification temperatures compared to Ni–Zn ferrite, Ni–Zn–Cu ferrites are promising candidates for use in producing multilayer type chip inductors [40-42].

Mn–Zn ferrites are used as a part of high flux transformers [43].

Small antennas are produced by winding a coil on ferrite used in transistor radio receivers. Ni–Mn–Co spinel ferrite T–DMB antennas were fabricated and characterized for antenna performance [39].

In computer non volatile memories application of ferrite materials provide an opportunity to demonstrate the resistive switching performance characteristics of a $Pt/NiFe_2O_4/Pt$ structure [44]. They store data even if the power supply goes off. Non–volatile memories are fixed up of ferrite materials as they are extremely stable against severe impact and vibrations.

$Ni_xZn_xFe_2O_4$ ferrites are used in microwave devices like circulators [45], isolators, switch phase shifters and in radar circuits [46].

$Mn_xNi_{0.5-x}Zn_{0.5}Fe_2O_4$ ferrites are used to decrease the eddy currents. Instead using non-conducting additives that locate preferentially on grain boundaries, Mn–Zn and Ni–Zn is used combine to synthesize such ferrites [47].

4.2 Biological applications

Magnetic materials in the form of nanoparticles, mainly magnetite (Fe_3O_4), are present in diverse living organisms [48] and can be practiced in a number of applications. Magnetic nanoparticles (MNP) can be synthesized in the laboratory by means of the well-known preparation methods. However, magnetic biogenic particles have more beneficial properties than synthetic ones; they have the definite size, breadth/length ratio, high chemical purity, they are perfect crystallographically, and sometimes they possess unusual crystallographic morphologies. Extracellular production of nanometer magnetite particles of several characters of bacteria has been reported [49]. In many instances, the biogenic particles retain a lipid layer which makes them very stable and easily biocompatible. Many biotechnological applications have been developed based on biogenic and synthetic magnetic micro and nanoparticles [50]. The magnetic nanoparticle has been utilized to guide radionuclide to specific tissues. An approach has been developed to directly label a radioisotope with ferrite particles [51] in in-vivo liver tissue in rats. Therapeutic applications are feasible by further conjugation with other health checks. In magnetic resonance imagery (MRI), magnetic superparamagnetic particles are selectively associated with healthy regions of some tissues (liver, for instance) [52]. Thermal energy from hysteresis loss of ferrites can be used in hyperthermia, that is, the heating of specific tissues or organs for the treatment of malignant neo-plastic disease. The temperature in tumor tissues rises and becomes more sensitive to radio and /or chemotherapy. In addition, of magnetite, several spinel ferrites (Me–Zn, with Me = Mn^{+2}, Co^{+2}, Fe^{+2}) are under investigation [53–54]. Today enzymes, antibodies, and other biologically active compounds can be immobilized in medical sciences. Such immobilized compounds can be targeted to a specific situation or can be removed from the organization by applying a magnetic field externally. The compounds can exert their action on the specific place or a tissue or can be used as affinity ligands to trap the cells or target molecules [55]. Magnetic nanoparticle can also be applied in a diversified of applications: modification, detection, isolation, and study of cells and isolation of biologically active compounds [56, 57]. According to the literature review, it is good to say there is a door opening for doing current research on spinel ferrites in association with the biological applications for human health care.

Conclusions

The structural properties of spinel ferrite is highly depended on the synthesis technique, heating temperature and time, amount of substitutions as well as chemical compositions. The distribution of cations in spinel ferrite plays an important role and hence magnetic properties of spinel ferrite are strongly depend upon the interaction between tetrahedral and octahedral cations. The bond length and bond angle of Metal-Oxygen-Metal also decides the magnetization along with other properties. The CFSE in octahedral and tetrahedral ligands with high spin and low spin and the superparamagnetism with magnetic anisotropy have been discussed.

References

[1]   H. S. Ahamad, N.S. Meshram, S. B. Bankar, S. J. Dhoble, K.G. Rewatkar, Structural properties of $Cu_xNi_{1-x}FeO_4$ nano ferrites prepared by urea gel microwave auto combustion method, Ferroelectrics, 516 (2017) 167–173. https://doi.org/10.1080/00150193.2017.1362285

[2] S. N. Sable, K. G. Rewatkar, V. M. Nanoti, Structural and magnetic behavioral improvisation of nanocalcium hexaferrites, Materials Science and Engineering B, 168 (2010) 156–160. https://doi.org/10.1016/j.mseb.2009.10.034

[3] C. Mamatha, M. Krishnaiah, C. S. Prakash, K. G. Rewatkar, Structural and electrical properties of aluminium substituted nano calcium ferrites, Procedia Materials Science, 5 (2014) 780–786. https://doi.org/10.1016/j.mspro.2014.07.328

[4] R. J. Hill, J. R. Craig, G.V. Gibbs, Systematics of the spinel structure type, International Journal of Physics and Chemistry Minerals, 4 (1979) 317-339. https://doi.org/10.1007/BF00307535

[5] A. D. Deshpande, K. G. Rewatkar, V. M. Nanoti, Study of morphology and magnetic properties of nanosized particles of zirconium – cobalt substituted calcium hexaferrites, Materials Today: Proceedings, 4 (2017) 12174–12179. https://doi.org/10.1016/j.matpr.2017.09.147

[6] J. N. Christy, K. G. Rewatkar, and P. S. Sawadh, Structural and magnetic behavior of M-type Co-Zr substituted calcium hexaferrites, Materials Today: Proceedings, 4 (2017) 11857–11865. https://doi.org/10.1016/j.matpr.2017.09.104

[7] B. Lavina, G. Salviulo, A. D. Giusta, Cation distribution and structure modelling of spinel solid solutions, Physics and Chemisty of Minerals, 29 (2002) 110–18. https://doi.org/10.1007/s002690100198

[8] L. Kumar, M. Kar, Influence of Al^{3+} ion concentration on the crystal structure and magnetic anisotropy of nanocrystalline spinel cobalt ferrite, Journal of Magnetism and Magnetic Materials, 323 (2011) 2042–2048. https://doi.org/10.1016/j.jmmm.2011.03.010

[9] G.D. Tang, Q.J. Han, J. Xu, D.H. Ji, Investigation of magnetic ordering and cation distribution in the spinel ferrites $Cr_xFe_{3-x}O_4$ ($0.0 \leq x \leq 1.0$), Physica B: Condensed Matter, 438 (2014) 91–96. https://doi.org/10.1016/j.physb.2014.01.010

[10] A. N. Birgani, M. Niyaifar, A. Hasanpour, Study of cation distribution of spinel zinc nano ferrite by X-ray, Journal of Magnetism and Magnetic Materials, 374 (2015) 179–181. https://doi.org/10.1016/j.jmmm.2014.07.066

[11] R.C. Kambale, N.R. Adhate, B. K. Chougule, Y. D. Kolekar, Magnetic and dielectric properties of mixed spinel Ni–Zn ferrite synthesized by citratenitrate combustion method, Journal of Alloys and Compounds, 491 (2010) 372–377. https://doi.org/10.1016/j.jallcom.2009.10.187

[12]   K. Sabri, A. Rais, K. Taibi, M. Moreau, B. Ouddane, A. Addou, Structural rietveld refinement and vibrational study of $MgCr_xFe_{2-x}O_4$, Physica B: Physics of Condensed Matter, 501 (2016) 38–44. https://doi.org/10.1016/j.physb.2016.08.011

[13]   H. M. Widatallaha, C. Johnsonb, F. J. Berryb, A. M. Gismelseed, Synthesis and cation distribution of copper–substituted spinel-related lithium ferrite, Journal of Physics and Chemistry of Solids, 67 (2006) 1817–1822. https://doi.org/10.1016/j.jpcs.2006.04.003

[14]   V.M. Nanoti, D.K. Kulkarni, Structural, electrical and magnetic study of the $Zn_{0.5}Ni_{0.5}Fe_xCr_{2-x}O_4$, Journal of Materials Science Letters, 15 (1996) 636–638. https://doi.org/10.1007/BF00579275

[15]   L. L. Lang, J. Xu, Z. Z. Li, G. D. Tang, Study of the magnetic structure and the cation distributions in MnCo, Physica B: Physics of Condensed Matter, 462 (2015) 47–53. https://doi.org/10.1016/j.physb.2015.01.008

[16]   P. Jadhav,K.Patankar, V. Mathe, N.L.Tarwal, Structural and magnetic properties of $Ni_{0.8}Co_{0.2-2x}Cu_xMn_xFe_2O_4$ spinel ferrites prepared via solution combustion route, Journal of Magnetism and Magnetic Materials, 385 (2015) 160–165. https://doi.org/10.1016/j.jmmm.2015.03.020

[17]   B. P. Jacob, A. Kumar, R. P. Pant, S. Singh, Influence of preparation method on structural and magnetic properties of nickel ferrite nanoparticles, Bulletin of Material Science, 34 (2011) 1345–1350. https://doi.org/10.1007/s12034-011-0326-7

[18]   J. P. Singh, G. Dixit, R.C. Srivastav, H. Kumara, Magnetic resonance in superparamagnetic zinc ferrite, Bulletin of Material Science, 36 (2013) 751–754. https://doi.org/10.1007/s12034-013-0528-2

[19]   X. Li, G. Wang, Low temperature synthesis and growth of superparamagnetic $Zn_{0.5}Ni_{0.5}Fe_2O_4$ nanosizedparticles, Journal of Magnetism and Magnetic Materials, (2009) 1276–128. https://doi.org/10.1016/j.jmmm.2008.11.006

[20]   M. Gharibshahiana, M.S.Nourbakhsha, Evaluation of superparamagnetic and biocompatible properties of mesoporous silica coated cobalt ferrite nanoparticles synthesized via microwave modified pechini method, Journal of Magnetism and Magnetic Materials, 425 (2017) 48–56. https://doi.org/10.1016/j.jmmm.2016.10.116

[21]   M.G. Naseri, M. Aara, E.B. Saion, Superparamagnetic magnesium ferrite nanoparticles fabricated by a simple thermal treatment method, Journal of

Magnetism and Magnetic Materials, 350 (2014) 141–147. https://doi.org/10.1016/j.jmmm.2013.08.032

[22] I. J. Bruvera, P. M. Zelis, M. P. Calatayud, G. F. Goya, F. H. Sanchez, Determination of the blocking temperature of magnetic nanoparticles, Journal of Applied Physics, 118 (2015) 184–304. https://doi.org/10.1063/1.4935484

[23] http://www.science20.com/mei/blog/blocking_temperature.

[24] Y. Tang, R. C. Flesch, T. Jin, A method for increasing the homogeneity of the temperature distribution during magnetic fluid hyperthermia with a Fe–Cr–Nb–B alloy in the presence of blood vessels, Journal of Magnetism and Magnetic Materials,432 (2017) 330–335. https://doi.org/10.1016/j.jmmm.2017.02.015

[25] Z. Abdel–Hamid, M. M. Rashad, S. M. Mahmoud, and A. T. Kandil, Electrochemical hydroxyapatite cobalt ferrite nanocomposite coatings as well hyperthermia treatment of cancer, Material Science and Engineering, 76 (2017) 827–838. https://doi.org/10.1016/j.msec.2017.03.126

[26] R. Colognato, A. Bonelli, D. Bonacchi, G. Baldi, L. Migliore, Analysis of cobalt ferrite nanoparticles induced genome, Journak of Nanotoxicology 1 (2007) 301–308. https://doi.org/10.1080/17435390701817359

[27] M. Coisson, G. Barrera, F. Celegato, L. Martino, S. N. Kane, S. Raghuvanshi, F. Vinai, P. Tiberto, Hysteresis losses and specific absorption rate measurements in magnetic nanoparticles for hyperthermia applications, Biochimica et Biophysica Acta (BBA), 1861 (2017) 1545–1558. https://doi.org/10.1016/j.bbagen.2016.12.006

[28] K. He, Y. Ma, B. Yang, C. Liang, X. Chen, C. Cai, The efficacy assessments of alkylating drugs induced by nano-Fe_3O_4/CA for curing breast and hepatic cancer, Spectrochimica Acta Part A: Molecular Spectroscopy, 173 (2017) 82–86. https://doi.org/10.1016/j.saa.2016.08.047

[29] S. Amiri, H. Shokrollahi, The role of cobalt ferrite magnetic nanoparticles in medical science, Journal of Material Science and Engineering, 33 (2013) 1–8. https://doi.org/10.1016/j.msec.2012.09.003

[30] N. N. Sarkar, N. S. Meshram, A. P. Bhat, K. G. Rewatkar, V. M. Nanoti, A review of nanoferrites: synthesis and application in hyperthermia, International Journal of Advanced Scientific and Technical Research, 5 (2015) 69–75.

[31] I. S. Oliveira, A. P. Guimafftes, A model for domain and domain wall NMR signals in magnetic materials, Journal of Magnetism and Magnetic Materials, 170 (1997) 277–284. https://doi.org/10.1016/S0304-8853(96)00723-8

[32] N. Singh, A. Agarwal, Effect of magnesium substitution on dielectric and magnetic properties of Ni Zn ferrite, Physica B Condensed Matter, 406 (2011) 687–692. https://doi.org/10.1016/j.physb.2010.11.087

[33] A. Ghasemi, Compositional dependence of magnetization reversal mechanism, magnetic interaction and Curie temperature of $Co_{1-x}Sr_xFe_2O_4$ spinel thin film, Journal of Alloys and Compounds, 645 (2015) 467–477. https://doi.org/10.1016/j.jallcom.2015.05.013

[34] A. J. Rondinone, A. C. Samia, Z. J. Zhang, Superparamagnetic relaxation and magnetic anisotropy energy distribution in $CoFe_2O_4$ spinel ferrite nanocrystallites, Journal of Physical Chemistry B, 103 (1999) 6876–6880. https://doi.org/10.1021/jp9912307

[35] V. Babayan, N.E. Kazantseva, Combined effect of demagnetizing field and induced magnetic anisotropy, Journal of Magnetism and Magnetic Materials, 324 (2012) 161–172. https://doi.org/10.1016/j.jmmm.2011.08.002

[36] L. Peleck, L. Diandra, Magnetic properties of nanostructured materials,Chemistry of Materials, 8 (1996) 1770–1783. https://doi.org/10.1021/cm960077f

[37] L.G. Antoshina, A.N. Goryaga, A.I. Kokorev, Magnetic anisotropy in ferrites spinels with frustrated magnetic structure, Journal of Magnetism and Magnetic Materials, 258 (2003) 516–519. https://doi.org/10.1016/S0304-8853(02)01130-7

[38] W. W. Porterfield, Inorganic chemistry, Academic Press, 2013

[39] S. Bae, Miniaturized broadband ferrite t–dmb antenna for mobile phone applications, IEEE Transactions on Magnetics, 46 (2010) 2361–2364. https://doi.org/10.1109/TMAG.2010.2044376

[40] H.V. Jamadar, M.B. Shelar, M.R. Bhandare, A.M. Shaikh, B. K. Chougule, Magnetic properties of nanocrystalline nickel zinc ferrites prepared by combustion synthesis, International Journal of Self Propagating High Temperature Synthesis, 20 (2011) 118–123. https://doi.org/10.3103/S1061386211020087

[41] T. Nakamura, Low-temperature sintering of Ni–Zn–Cu ferrite and its permeability spectra, Journal of Magnetism and Magnetic Materials, 168 (1997) 285–291. https://doi.org/10.1016/S0304-8853(96)00709-3

[42] J. Y. Hsu, W. S. Ko, H. D. Shen, C. J. Chen, Low temperature fired Ni–Cu–Zn ferrite, IEEE Transactions on Magnetics, 30 (1994) 4875–4877. https://doi.org/10.1109/20.334251

[43] K. Padmanabhan, Electronic components. Firewall Media, 2006.

[44]   W. Hu, N. Qin, G. Wu, Y. Lin, S. Li, D. Bao, Opportunity of spinel ferrite materials in nonvolatile memory device applications based on their resistive switching performances,Journal of the American Chemical Society, 134 (2012) 14658–14661. https://doi.org/10.1021/ja305681n

[45]   M. Wu, A. Hoffmann,E. Robert, C. Robert, L. Stamps, Recent Advances in Magnetic Insulators–From Spintronics to Microwave Applications, first ed., Academic Press, 64 (2013).

[46]   R. Valenzuela, Novel applications of ferrites, Physics Research International, 2012 (2011) 1–9. https://doi.org/10.1155/2012/591839

[47]   A. Verma, M.I. Alam, R. Chatterjee, T.C. Goel, R. G. Mendiratta, Development of a new soft ferrite core for power applications, Journal of Magnetism and Magnetic Materials, 300 (2006) 500–505. https://doi.org/10.1016/j.jmmm.2005.05.040

[48]   A. R. West, Solid state chemistry and its applications, Second edition, Wiley, 2014.

[49]   S. Venkateswarlu, Y. S. Rao, T. Balaji, B. Prathima, Biogenicsyn thesis of Fe_3O_4 magnetic nanoparticles using plantain peel extract, Journal of Materials Letters, 100 (2013) 241–244.

[50]   M. Ravichandran, G. Oza, S. Velumani, Onedimensional ordered growth of magnetocrystalline, Journal of Materials Letters, 135 (2014) 67–70. https://doi.org/10.1016/j.matlet.2014.07.154

[51]   K. Iwahori, J. Watanabe, Y. Tani, H. Seyama, N. Miyata, Removal of heavy metal cations by biogenic magnetite nanoparticles produced, Journal of Bioscience and Bioengineering, 117 (2014) 333–335. https://doi.org/10.1016/j.jbiosc.2013.08.013

[52]   E.C. Sadasiv, T.T. Yeh, P.W. Chang, Protein pI alteration related to strain variation of infectious bronchitis virus, an avian Coronavirus, Journal of Virological Methods, 33 (1991) 115–125. https://doi.org/10.1016/0166-0934(91)90012-O

[53]   A. Jorda, R. Scholz, K. M. Hauff, Presentation of a new magnetic field therapy system, Journal of Magnetism and Magnetic Materials, 225 (2001) 118–126. https://doi.org/10.1016/S0304-8853(00)01239-7

[54]   A. Hooda, S. Sanghi, A. Agarwal, R. Dahiya, Crystal structure refinement, dielectric and magnetic properties of Ca/Pb substituted $SrFe_{12}O_{19}$ hexaferrites, Journal of Magnetism and Magnetic Materials, 387 (2015) 46–52. https://doi.org/10.1016/j.jmmm.2015.03.078

[55]   C. M. Ahameda, M. J. Akhtar, H. A. Alhadlaqa, A. Alshamsan, Copper ferrite nanoparticle induced cytotoxicity and oxidative stressin human breast cancer, Colloids and Surfaces B: Biointerfaces, 142 (2016) 46–54. https://doi.org/10.1016/j.colsurfb.2016.02.043

[56]   K. Kobayashi, S. Sagae, T. Takeda, M. Sugimura, Y. Nishioka, Genetic analysis of familial and multiple malignancies, International Journal of Gynecology and Obstetrics, 66 (1999) 149–153. https://doi.org/10.1016/S0020-7292(99)00066-1

[57]   A. Cvetanovic, J. S. Gajic, P. Maskovic, S. Savic, Antioxidant and biological activity of chamomile extracts obtained by different techniques perspective of using superheated water forisolation of biologically active compounds, Industrial Crops and Products, 65 (2015) 582–591. https://doi.org/10.1016/j.indcrop.2014.09.044

Keyword Index

About the Editors

Prof. Rajshree B. Jotania

Dr. Rajshree B. Jotania is a professor of Physics, Department of Physics, Electronics and space science, University School of Sciences at Gujarat University, Ahmedabad, India. She obtained her B.Sc., M.Sc. and Ph.D from Saurashtra University, Rajkot, India. She was Junior Research Fellow (DAE-BRNS project) during 1987 to 1989 at Physics Department, Saurashtra University, Rajkot, India. She obtained a few regional, and national awards for contribution toward scientific research. She worked at National Chemical Laboratory, Pune, India for two months as a Summer Visiting Teacher Fellow in 2005 and as a Visiting Scientist fellow in 2011. She possesses total 28 years of teaching experience at UG and PG level. She is a member of Board of studies at few Universities of Gujarat, India and a Mentor of DST-INSPIRE (Department of Science and Technology- Innovation in Science Pursuit for Inspired Research) program. She has published more than 80 papers in various research journals and conference proceedings. She has delivered more than 20 invited talks at various DST-INSPIRE Internship science camp in India. She has edited two books entitled 'Ferrites and ceramic composites' (Vol. I & II, Trans Tech Publisher, Switzerland). She has visited Singapore, Malaysia and New York for presenting Research work. She has attended more than 40 international, national conferences/symposiums/seminars/ academy meeting and worked as a chair person in a few international and national conferences. She possesses a life membership of eight professional bodies and she has guided four Ph. D, eight M. Phil students. To date she has completed five research projects of various agencies.

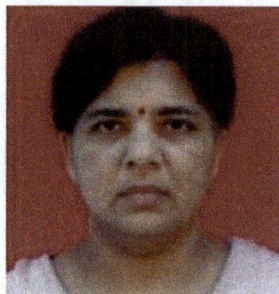

Prof. Sami H. Mahmood

Currently, Professor Sami Mahmood is a professor of Physics, and Dean of School of Science at the University of Jordan, Amman. He obtained his B.Sc. in Physics from The University of Jordan, Amman in 1978, and Ph.D in Physics from Michigan State University, East Lansing, Michigan in 1986. He assumed several administrative positions at different Jordanian Universities such as Director of the Center for Theoretical and Applied Physical Sciences (1992-1994), Chairman of Physics (1996-1998), Dean of Science (1998-2004), Dean of Scientific Research and Graduate Studies (2007-2009), Director of the Center of Accreditation and Quality Assurance, and Vice President (2004-2005). He obtained several national, regional and international Awards and Honors for Academic excellence and contribution to science. To date he has published over 100 articles in peer reviewed journals and conference proceedings. Other academic and scientific activities included membership in the editorial boards of several national and international research journals, participation and management of nationally and internationally funded projects concerned with the establishment of new academic programs, capacity building, and program development, in addition to his active participation in scientific committees and councils of Scientific Research Funds in Jordan, and organization of national and international conferences.